MR Enterography

Silvio Mazziotti • Alfredo Blandino
Giuseppe Cicero
Editors

MR Enterography

Crohn's Disease and Beyond

Second Edition

Springer

Editors
Silvio Mazziotti
Department of Biomedical Sciences
University of Messina
Messina, Italy

Alfredo Blandino
Department of Biomedical Sciences
University of Messina
Messina, Italy

Giuseppe Cicero
Department of Biomedical Sciences
University of Messina
Messina, Italy

ISBN 978-3-031-11932-3   ISBN 978-3-031-11930-9 (eBook)
https://doi.org/10.1007/978-3-031-11930-9

This Springer imprint is published by the registered company Springer Nature Switzerland AG
The registered company address is: Gewerbestrasse 11, 6330 Cham, Switzerland

Contents

Contributors

Giorgio Ascenti Department of Biomedical Sciences and Morphological and Functional Imaging, University Hospital Messina, Messina, Italy

Alfredo Blandino Department of Biomedical Sciences and Morphological and Functional Imaging, University Hospital Messina, Messina, Italy

Christian Booz Division of Experimental Imaging, Department of Diagnostic and Interventional Radiology, University Hospital Frankfurt, Frankfurt am Main, Germany

Giuseppe Cicero Department of Biomedical Sciences and Morphological and Functional Imaging, University Hospital Messina, Messina, Italy

Tommaso D'Angelo Department of Biomedical Sciences and Morphological and Functional Imaging, University Hospital Messina, Messina, Italy

Simon S. Martin Department of Diagnostic and Interventional Radiology, University Hospital Frankfurt, Frankfurt am Main, Germany

Division of Experimental Imaging, Department of Diagnostic and Interventional Radiology, University Hospital Frankfurt, Frankfurt am Main, Germany

Silvio Mazziotti Department of Biomedical Sciences and Morphological and Functional Imaging, University Hospital Messina, Messina, Italy

Thomas J. Vogl Department of Diagnostic and Interventional Radiology, University Hospital Frankfurt, Frankfurt am Main, Germany

Ibrahim Yel Department of Diagnostic and Interventional Radiology, University Hospital Frankfurt, Frankfurt am Main, Germany

Division of Experimental Imaging, Department of Diagnostic and Interventional Radiology, University Hospital Frankfurt, Frankfurt am Main, Germany

1 Introduction

Silvio Mazziotti, Alfredo Blandino,
and Giuseppe Cicero

Gastrointestinal imaging (GI) has more than one hundred and twenty years of history, since the first clinical application of X-ray within the gut dates back to 1897 [1].

Over the following years, GI has been essentially based on the radiographic and fluoroscopic approaches till the introduction of cross-sectional modalities, such as Ultrasound (US), Computed-Tomography (CT) and Magnetic Resonance (MR) [2].

However, within these imaging modalities, specific techniques aimed to the evaluation of intestinal loops started to emerge only in the last 25 years [3–6].

This evolution has improved the assessment of a number of benign and malignant intestinal conditions.

Above all, a steady increase in accuracy has been recorded for the assessment of Inflammatory Bowel diseases (IBDs).

IBDs, namely Crohn's disease (CD) and Ulcerative Colitis (UC), represent chronic pathologic conditions affecting the small and/or the large bowel.

While the precise etiology is still unknown, the pathogenesis consists in an abnormal immune response against the intestinal microbiota resulting from a combination of genetic predisposition and environmental risk factors [7, 8].

Symptoms are characterized by an alternation of remission and recurrence periods and are grossly represented by abdominal pain, diarrhea with or without bleeding, weight loss, and fever [7, 8].

The clinical picture between the two is different, since CD determines a transmural involvement and can affect any part of the gastrointestinal tract, from the mouth to the anus, whereas UC mainly affects the mucosal layer of the large bowel with a progressive distal to proximal involvement [7, 9, 10].

Incidence of IBDs is significantly increased over the last twenty years [11].

Although it was previously considered a prerogative of Europe and North America, high rates are now reported also in non-Western countries on both pediatric and adult subjects, with an estimated population of more than 20 million IBD patients all over the world [9, 10, 12–14].

The final diagnosis is reached through the combination of clinical symptoms, laboratory tests, and endoscopy with biopsy.

On the radiological side, imaging supports the diagnosis and affords useful information about disease extension and features.

Currently, thanks to the constant technological progress, several imaging modalities are available for evaluation of IBD patients, each of them provided with strengths points and limitations [15].

S. Mazziotti (✉) · A. Blandino · G. Cicero
Department of Biomedical Sciences and Morphological and Functional Imaging, University Hospital Messina, Messina, Italy
e-mail: smazziotti@unime.it; ablandino@unime.it; gcicero@unime.it

S. Mazziotti et al. (eds.), *MR Enterography*, https://doi.org/10.1007/978-3-031-11930-9_1

Therefore, the diagnostic choice should be based on the patient's characteristics and clinical needs.

In particular, Magnetic Resonance Enterography (MRE) is widely considered the imaging modality of choice, taking advantage from the comprehensive evaluation of the small bowel as well as the abdominal cavity, the high contrast resolution of soft tissues, and the lack of ionizing radiation [16].

The aim of this volume is therefore to deepen and share knowledge about the current state of the art of MR enterography in CD patients. However, this work goes even beyond, addressing additional topics, such as the remaining imaging techniques employable in the intestinal district, including the study of the anal canal, and the discussion of clinical scenarios other than IBDs [17–20].

References

1. Lindemann E. Demonstration of roentgen pictures of the normal and distended stomach. Dtsch Med Wochenschr. 1897;23:266–7.
2. Eisenberg RL, Margulis AR. Brief history of gastrointestinal radiology. Radiographics. 1991;11(1):121–32. https://doi.org/10.1148/radiographics.11.1.1996385. PMID: 1996385
3. Raptopoulos V, Schwartz RK, McNicholas MM, Movson J, Pearlman J, Joffe N. Multiplanar helical CT enterography in patients with Crohn's disease. AJR Am J Roentgenol. 1997;169:1545–50.
4. Pallotta N, Baccini F, Corazziari E. Ultrasonography of the small bowel after oral administration of anechoic contrast solution. Lancet. 1999;353:985–6.
5. Faber SC, Stehling MK, Holzknecht N, Gauger J, Helmberger T, Reiser M. Pathologic conditions in the small bowel: findings at fat-suppressed gadolinium-enhanced MR imaging with an optimized suspension of oral magnetic particles. Radiology. 1997;205(1):278–82. https://doi.org/10.1148/radiology.205.1.9315000.
6. Aschoff AJ, Zeitler H, Merkle EM, Reinshagen M, Brambs HJ, Rieber A. MR-Enteroklyse zur kernspintomographischen Diagnostik entzündlicher Darmerkrankungen mit verbesserter Darmkontrastierung [MR enteroclysis for nuclear spin tomographic diagnosis of inflammatory bowel diseases with contrast enhancement]. Rofo. 1997;167(4):387–91. German. https://doi.org/10.1055/s-2007-1015549.
7. Wallace KL, Zheng LB, Kanazawa Y, Shih DQ. Immunopathology of inflammatory bowel disease. World J Gastroenterol. 2014;20(1):6–21. https://doi.org/10.3748/wjg.v20.i1.6. PMID: 24415853; PMCID: PMC3886033
8. Kobayashi T, Siegmund B, Le Berre C, Wei SC, Ferrante M, Shen B, Bernstein CN, Danese S, Peyrin-Biroulet L, Hibi T. Ulcerative colitis. Nat Rev Dis Primers. 2020;6(1):74. https://doi.org/10.1038/s41572-020-0205-x.
9. Goodman WA, Erkkila IP, Pizarro TT. Sex matters: impact on pathogenesis, presentation and treatment of inflammatory bowel disease. Nat Rev Gastroenterol Hepatol. 2020;17(12):740–54. https://doi.org/10.1038/s41575-020-0354-0. Epub 2020 Sep 8. PMID: 32901108; PMCID: PMC7750031
10. Roda G, Chien Ng S, Kotze PG, Argollo M, Panaccione R, Spinelli A, Kaser A, Peyrin-Biroulet L, Danese S. Crohn's disease. Nat Rev Dis Primers. 2020;6(1):22. https://doi.org/10.1038/s41572-020-0156-2. Erratum in: Nat Rev Dis Primers. 2020 Apr 6;6(1):26. Erratum in: Nat Rev Dis Primers. 2020 May 20;6(1):42. Erratum in: Nat Rev Dis Primers. 2020 Jun 19;6(1):51
11. GBD 2017 Inflammatory Bowel Disease Collaborators. The global, regional, and national burden of inflammatory bowel disease in 195 countries and territories, 1990–2017: a systematic analysis for the Global Burden of Disease Study 2017. Lancet Gastroenterol Hepatol. 2020;5(1):17–30. https://doi.org/10.1016/S2468-1253(19)30333-4. Epub 2019 Oct 21. PMID: 31648971; PMCID: PMC7026709
12. Ng SC, Shi HY, Hamidi N, Underwood FE, Tang W, Benchimol EI, Panaccione R, Ghosh S, Wu JCY, Chan FKL, Sung JJY, Kaplan GG. Worldwide incidence and prevalence of inflammatory bowel disease in the 21st century: a systematic review of population-based studies. Lancet. 2017;390(10114):2769–78. https://doi.org/10.1016/S0140-6736(17)32448-0. Epub 2017 Oct 16. Erratum in: Lancet. 2020 Oct 3;396(10256):e56
13. Huang JG, Aw MM. Pediatric inflammatory bowel disease in Asia: epidemiology and natural history. Pediatr Neonatol. 2020;61(3):263–71. https://doi.org/10.1016/j.pedneo.2019.12.008. Epub 2019 Dec 27
14. Gasparetto M, Guariso G. Highlights in IBD epidemiology and its natural history in the paediatric age. Gastroenterol Res Pract. 2013;2013:829040. https://doi.org/10.1155/2013/829040. Epub 2013 Dec 24. PMID: 24454343; PMCID: PMC3884601
15. Cicero G, Mazziotti S. Crohn's disease at radiological imaging: focus on techniques and intestinal tract. Intest Res. 2021;19(4):365–78. https://doi.org/10.5217/ir.2020.00097. Epub 2020 Nov 25. PMID: 33232590; PMCID: PMC8566824
16. Mazziotti S, Ascenti G, Scribano E, Gaeta M, Pandolfo A, Bombaci F, Donato R, Fries W, Blandino A. Guide to magnetic resonance in Crohn's disease: from common findings to the more rare complicances.

Inflamm Bowel Dis. 2011;17(5):1209–22. https://doi.org/10.1002/ibd.21548. Epub 2010 Nov 5
17. Cicero G, Blandino A, D'Angelo T, Booz C, Vogl TJ, Ascenti G, Mazziotti S. Mimicking conditions of intestinal Crohn's disease: magnetic resonance enterography findings. Jpn J Radiol. 2022;40(1):19–28. https://doi.org/10.1007/s11604-021-01177-7. Epub 2021 Jul 25
18. Amzallag-Bellenger E, Oudjit A, Ruiz A, Cadiot G, Soyer PA, Hoeffel CC. Effectiveness of MR enterography for the assessment of small-bowel diseases beyond Crohn disease. Radiographics. 2012;32(5):1423–44. https://doi.org/10.1148/rg.325115088.
19. Cicero G, Ascenti G, Blandino A, Pallio S, Abate C, D'Angelo T, Mazziotti S. Magnetic resonance imaging of the anal region: clinical applications. J Clin Imaging Sci. 2020;10:76. https://doi.org/10.25259/JCIS_180_2020. PMID: 33274120; PMCID: PMC7708963
20. Cicero G, Ascenti G, Bottari A, Catanzariti F, Blandino A, Mazziotti S. MR enterography: what is next after Crohn's disease? Jpn J Radiol. 2019;37(7):511–7. https://doi.org/10.1007/s11604-019-00838-y. Epub 2019 Apr 9

Small Bowel Imaging Other Than MR-Enterography

2

Alfredo Blandino, Thomas J. Vogl, Simon S. Martin, Ibrahim Yel, and Christian Booz

2.1 X-Ray and Fluoroscopy

Abdominal X-ray plain films may represent the first step in the imaging framework of CD patients, especially in case of emergency.

They can be obtained on erect and supine views, which can be further complemented by supine lateral or lateral decubitus projections.

X-ray only provides general information about bowel gas pattern, including quantitative assessment of intestinal distention, distribution of dilated loops with localization of obstruction and recognition of pneumoperitoneum.

After contraindications are excluded (i.e., perforation or toxic megacolon), fluoroscopic exam may be conducted [1].

On the other hand, fluoroscopy mainly embraces two different techniques for small-bowel assessment, including enteroclysis (small-bowel enteroclysis, SBE) and follow-through (small bowel follow-through, SBFT), which historically represented the standard approaches for CD patients until the advent of cross-sectional imaging [2–4].

Both the techniques require intestinal distention and opacification through an enteric contrast agent (approximately 700–750 mL), but differ for the way of its administration: while at SBFT the patient is asked to drink, SBE requires the positioning of a nasojejunal tube [5].

Due to its invasiveness and discomfort, the latter approach is mainly reserved to non-compliant patients, such as pediatrics.

For what concern the intraluminal contrast medium, barium sulfate water solution is generally preferred since its viscosity allows an accurate coating of the intestinal mucosal layer and it has an inert character to the lungs if aspiration occurs during the exam.

However, clinical suspicion of visceral perforation is an absolute contraindication to barium administration as it cannot be absorbed by the peritoneum and its presence inside the abdominal cavity leads to exudation of extracellular fluid

A. Blandino (✉)
Department of Biomedical Sciences and Morphological and Functional Imaging, University Hospital Messina, Messina, Italy
e-mail: ablandino@unime.it

T. J. Vogl
Department of Diagnostic and Interventional Radiology, University Hospital Frankfurt, Frankfurt am Main, Germany
e-mail: t.vogl@em.uni-frankfurt.de

S. S. Martin · I. Yel
Department of Diagnostic and Interventional Radiology, University Hospital Frankfurt, Frankfurt am Main, Germany

Division of Experimental Imaging, Department of Diagnostic and Interventional Radiology, University Hospital Frankfurt, Frankfurt am Main, Germany
e-mail: simon.martin@kgu.de; Ibrahim.Yel@kgu.de

C. Booz
Division of Experimental Imaging, Department of Diagnostic and Interventional Radiology, University Hospital Frankfurt, Frankfurt am Main, Germany
e-mail: christian.booz@kgu.de

S. Mazziotti et al. (eds.), *MR Enterography*, https://doi.org/10.1007/978-3-031-11930-9_2

and albumin resulting in hypoproteinemia with consequent hypovolemia and fibroplastic proliferation associated to adhesions and possible onset of peritonitis. In this case, water-soluble iodine-based agents can be used [1].

After the administration of the enteric agent, radiographic series are acquired to document its progression at least until the caecum is reached [6].

The evaluation of terminal ileum may be also improved by views during compression [7].

Early radiological findings of CD include irregularity of the intervillous spaces, thickening of valvulae conniventes due to inflammatory infiltration and aphthoid mucosal erosions [1, 6].

On the late stages of the disease, ulcerations may be detected as linear "contrast-plus" images within the mural thickness. If ulcers are surrounded by edematous mucosa (forming "pseudopolyps"), the typical "cobblestone" appearance can be appreciated [1, 7].

Ulcers may also develop into fistulas and sinus-tracts, showing extra-parietal extension.

An additional sign of inflammation can be represented by the abnormal separation of bowel loops due to inflammatory mesenteric infiltration (the so-called "creeping fat" sign) (Fig. 2.1) [1, 7].

Fibrotic pattern of CD can be suspected due to the presence of strictures as well as unusual attraction of small-bowel loops [1].

Water-soluble contrast medium can be also injected in a retrograde way using an external catheter through ileostomy in order to evaluate possible obstruction or enterocutaneous fistulae for assessment of enterocutaneous fistulae ("fistulography") [1, 6].

Barium colonic enema represents an option for evaluation of the large bowel [1, 6].

The main advantage of fluoroscopic techniques consists in the ongoing appraisal of contrast medium progression with simultaneous assessment of normal or impaired peristalsis [4, 7].

On the other hand, the major limit relies on the transmural and extraluminal assessment [3, 8].

With the worldwide spread of cross-sectional imaging, the recourse to fluoroscopic techniques is currently limited and it is related to simplicity of their performance, low healthcare costs, easy access rather than CT-scan or MRI schedules [5, 8, 9]. **e.**

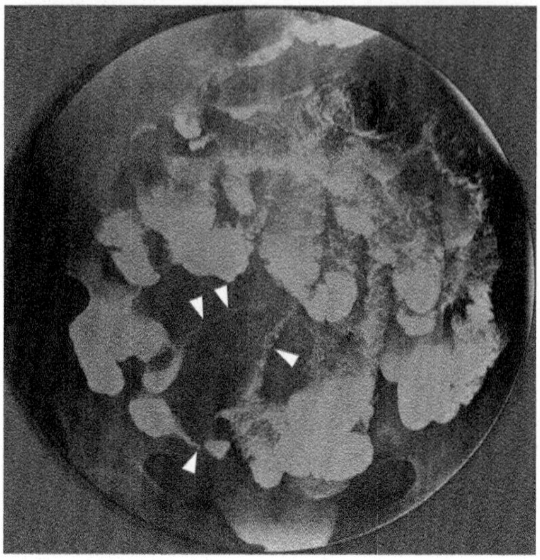

Fig. 2.1 Small bowel follow-through performed during assumption of a barium-based contrast material in a CD patient. Multiple strictures of ileal loops are detectable within the lower right abdominal quadrant (arrowheads). The ileal loops are distanced, an indirect sign of mesenteric fat tissue hypertrophy subsequent to chronic inflammation

Over CT and MRI, SBE and SBFT can afford more accurate mucosal detail, however with lower sensitivity in detecting active inflammation and delivering poor information about extra-intestinal involvement.

Moreover, considering its two-dimensional appraisal, overlapping loops may hide intestinal pathology, especially when deeply located (i.e., pelvic cavity) [3, 9].

Another drawback is represented by the delivered radiation dose, considering that the exposure may significantly vary due to the floating examination time according to peristalsis [2, 5, 8].

2.2 Ultrasound

Ultrasound (US) is a reliable imaging modality for evaluation of intestinal involvement of CD patient.

Its main benefits consist in wide availability, noninvasiveness, low healthcare costs, and lack of ionizing radiation.

The latter is a significant advantage considering the wide onset of CD in young patients and the need of repeating imaging examinations during their lifetime.

Moreover, US provides a real-time appraisal of intestinal peristalsis and can be used as a guide in case of interventional procedures, such as abscess drainage [10–12].

The basic US approach is represented by greyscale B-mode evaluation without any preparation, apart from previous fasting (4–6 h).

The exam should begin with a panoramic view of the small and the large bowel using a 3–5 MHz convex array probe, starting in the right iliac fossa or by exploring the large bowel.

Compression and rotation of the probe on different axes can improve detection of the affected loops [11].

Once the pathology is identified, a high frequency linear or microconvex probe (up to 18 MHz), owing to the higher spatial resolution, can be used for a further characterization of the bowel wall [12, 13].

The bowel wall is normally displayed as an alternation of hyper- and hypoechoic layers.

Starting from the innermost, the first one (hyperechoic) corresponds to the interface between the lumen and the superficial mucosa, the second (hypoechoic) to the deep mucosa, the third (hyperechoic) to the submucosa, the fourth (hypoechoic) to the muscularis propria, the fifth (hyperechoic) to the serosa [12, 14].

Considering the inflammatory nature of bowel wall thickening in CD, B-mode US can be easily complemented by information about vascular flow using color- and/or power-Doppler.

Increased vascular signals at color-Doppler can be expressed through semiquantitative scores, such as the Limberg's one, which however are subjective and depends on the operator's expertise [15].

A quantitative method can be obtained using pulse-wave Doppler, especially of the superior mesenteric artery and calculating the resistive index (RI), which has demonstrated encouraging results in differentiating inflammatory (low RI) from malignant (high RI) neovascularization [14].

The main US findings in CD are represented by increased mural thickness (>3 mm), stenosis in terms of lumen narrowing with upstream loop dilation (at least 2.5/3 cm), alteration of wall stratification which can determine the appearance of a "target sign" on transverse view due to pronounced hypoechoic edema [11, 12, 14, 15].

The affected intestinal segment may appear rigid at compression, with decreased peristaltic activity, and hypervascularized at color-Doppler [11, 14].

Accessory signs of CD are represented by augmented echogenicity and hypertrophy of the mesenteric fat tissue and perivisceral lymphadenopathies [11, 12].

Within the wall width, ulcerations may be also detected as linear hyperechoic tracts which can also show an extra-parietal development forming sinus tracts or fistulae [11, 12].

Fistulae may be recognized by the presence of air bubbles or increased vascular flow at color-Doppler.

Involvement of the perivisceral fat tissue may also lead to phlegmons and abscesses development.

While the former can be detected as irregular hypoechoic "mass-like" structures with central blood-flow signals at color-Doppler, abscesses appear as inhomogeneous fluid collections with gas bubbles and peripheral vascular signs [11, 12].

Limitations of US include dependency on operator's experience and incomplete visualization of the whole bowel, especially jejunal and deep pelvic loops including rectum, with possible underestimation of disease extent [11, 15].

Variability among operators may lead to differences in quantification of mural thickening as well as disease localization for small-bowel loops other than terminal ileum [12, 15].

Over the last years, different techniques have been developed in order to further increase the diagnostic accuracy of US in CD evaluation, including contrast-enhanced US (CEUS), Small Intestine Contrast Ultrasonography (SICUS), and US Elastography (USE).

CEUS is relatively novel technique that allows a real-time assessment of microvasculature by exploiting the intravenous administration of a second-generation microbubble contrast agent (SonoVue, SV, Bracco, Italy) [11, 12].

While color-Doppler is able to detect vascular signals from larger blood vessels that supply the bowel walls, CEUS can evaluate intramural microcirculation and inflammatory neoangiogenesis as well as mesenteric vascularization (the so-called "comb sign") [16].

Promising results are now emerging from the differentiation of active inflammation from fibrosis, since the former is characterized by hyperemia with consequent intense contrast enhancement [11, 14, 15, 17, 18].

In this sense, CEUS has demonstrated a good to excellent concordance with MRE findings and endoscopic features [11].

However, since both the phenomena are often coexistent, a clear distinction is often challenging [11].

CEUS has also shown improved sensitivity and specificity than standard US for diagnosis of CD as well as increased sensitivity in detection of postoperative recurrences [13, 15].

Quantification of bowel wall vascularity can be also employed for monitoring of disease activity after medical therapy or for recurrence detection after surgery in CD [16].

CEUS can be hampered by peristaltic movements. Furthermore, it allows evaluating only one owel loops at a time [15].

Another already consolidated application of CEUS in CD patients is represented by the differentiation of abscesses, characterized only by parietal enhancement, from phlegmons, that show intralesional enhancement [11, 14, 16, 17].

The performance of SICUS requires the ingestion of a variable dose (125–800 mL) of water solution of polyethylene glycol (PEG). The intraluminal dilation allows a real-time assessment of

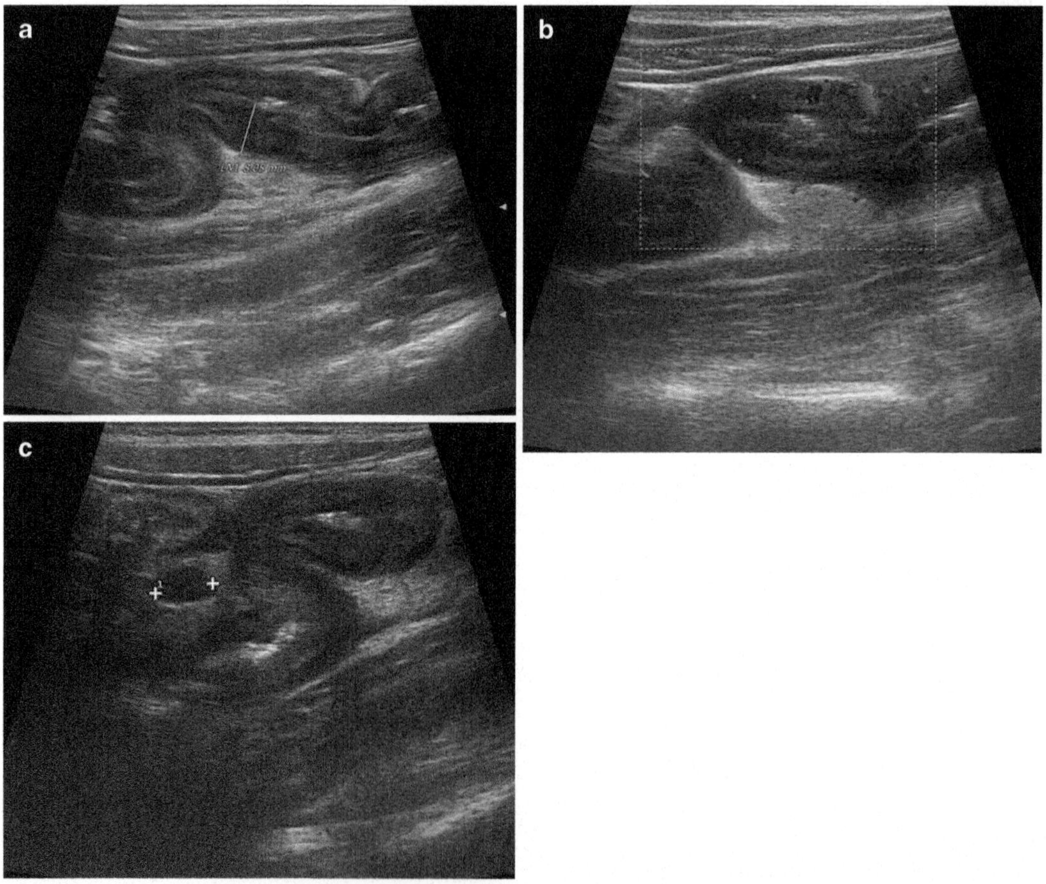

Fig. 2.2 Small Intestine Contrast Ultrasonography (SICUS) performed on a CD patient. Wall thickening of the last ileal loop with luminal narrowing was detected (**a**) together with increased vascular signals at color-Doppler due to inflammation (**b**). An enlarged lymph node was also identifiable within the surrounding adipose tissue (**c**)

peristalsis and a more detailed characterization of bowel loops due to the avoidance of intraluminal gas artifacts (Fig. 2.2) [14, 15, 17, 18].

Different studies confirmed the higher detection rate of intestinal pathologic segments than conventional US, especially in case of strictures and improved sensitivity for postoperative recurrences [13].

Despite the undoubted benefits, SICUS still remains an underperformed technique, mainly due to the operator experience required and the long exam duration, that can last up to 60 minutes [14, 15, 17, 18].

Elastography is a technique whose use has been largely employed for lesion characterization on thyroid, breast, and liver and it is based on evaluation of tissue stiffness.

Elastography includes two main techniques: "shear wave elastography" (SWE), which provides quantitative parameters from measurement of shear wave velocity, and "strain elastography" (SE), that displays as a colorimetric map the stiffness differences between a targeted tissue and the surroundings after application of an external pressure [16, 19].

Therefore, elastography could be exploited for stricture characterization in CD, differentiating active inflammation from fibrosis, since the latter is characterized by higher stiffness [15, 19].

However, up to now, quantitative and standardized measurements are lacking.

Although the diagnostic accuracy seems to increase when complemented by CEUS, the performance of both the elastographic techniques remains currently limited due to operator-dependency that implies a subjective evaluation with consequent low reproducibility [16, 17, 19].

2.3 Computed Tomography

Computed Tomography Enterography (CTE) in the evaluation of CD patients comes with many advantages including widespread availability, comprehensive evaluation of the whole abdominal cavity, faster scan times, and lower healthcare costs than MR [20, 21].

CTE requires the administration of both oral and intravenous contrast media in order to maximize distinction between the lumen and intestinal wall enhancement [22].

The enteric contrast agent can be orally assumed by the patient (the actual CTE) or through a nasojejunal tube (named CT-Enteroclysis). The former is usually preferred due to its noninvasiveness [21, 23].

An adequate distension of small-bowel loops is mandatory and can be obtained by the use of a large volume of oral contrast material (450 mL to 2 L, according to the patient weight) over a relatively short period of time (45–60 min) in order to avoid intestinal absorption of water component [20, 24].

Neutral or negative enteric contrast agents, such as low density barium sulfate suspensions, mannitol, or polyethylene glycol (PEG), are usually preferred since they consent a low-attenuation distension of the intestinal loops without any risk of covering wall enhancement after intravenous contrast medium injection [25–27].

Assumption of spasmolytic agents in order to avoid peristaltic artifacts can be performed before scanning [27].

Abdominal and pelvic cavities have to be included in the field of view as well as the anal canal and the perianal region [20, 24].

The exam starts with an unenhanced scan after which the intravenous iodinated contrast medium is injected according to the iodine concentration and a weight-based algorithm (1.6 mL/kg), at a rate of 3–4 mL/s and followed by 50 mL of saline solution [24, 27, 28].

Afterwards, a multiphasic contrast-enhanced study is performed, generally obtaining an arterial phase at 35–45 s after the injection starts, an enteric phase at 50–70s, and a delayed phase at 90s [20, 24].

CTE has demonstrated high sensitivity and specificity rates for small-bowel CD detection, non-inferior to MRI [23].

Radiological findings of active inflammation are represented by mural thickening (>3 mm) characterized by contrast enhancement, which can appear homogeneous, limited to the mucosal side or layered, due to intramural edema interposed between mucosal and serosal hyperenhancement [21, 28].

On the other hand, intramural fat infiltration is more indicative of a chronic process, whereas lack or delayed enhancement and loss of stratification are usually associated to fibrosis [21–28].

The penetrating pattern of the disease can be appreciated by visualizing transmural ulcerations that can develop into periluminal phlegmons, fistulae, or abscesses [22, 27].

Strictures are identified as luminal narrowing with upstream loop dilation and can be caused by inflammation, fibrosis, or a combination of the two phenomena [20, 27].

Extravisceral signs consist of mesenteric fat stranding due to edema, engorged vasa recta ("comb sign"), and enlarged lymph nodes [21, 22, 24, 27, 28].

Fibro-fatty proliferation of perivisceral mesentery can be appreciated as an abnormal loop separation on contrast examination [21, 27].

Post-processing reconstruction, such as maximum intensity projection (MIP), may be useful for an intestinal overview and a mesenteric vessels assessment [20, 21].

Moreover, CT-scan has the spatial and contrast resolution for detecting extra-intestinal alterations due to metabolic changes related to CD, such as biliary and kidney stones [27].

Pitfalls on CTE assessment of the small bowel can derive from insufficient loop dilation or peristaltic spasms. In this cases, repeated scans can help in distinguishing real strictures, tough at the cost of exposing the patient to a higher radiation dose [21, 27].

In this context, further possibilities come from dual-energy CT technology (DECT), which exploits post-processing reconstructions obtained at different energy levels.

Among the various DECT tools, Virtual Monochromatic Images (VMI) permit to obtain images at a desired energy level, as the body tissues were hit by a perfectly monoenergetic radiation beam. Therefore, lowering the energy level close to the iodine k-edge allows to emphasize contrast enhancement [29–31].

DECT also offers the opportunity of performing iodine density measurements. This allows a real contrast enhancement quantification in terms of iodine concentration, overcoming a simple tissue density measurement which could be impaired by partial volume artifacts (Fig. 2.3) [32, 33].

DECT tools have a great diagnostic potential and could be of great help within the next future for disease activity estimation, response to therapy assessment and inflammation from fibrosis distinction [32].

The major concern about the use of CTE is obviously related to the radiation exposure due to the number of radiological exams required due to the relapsing-remitting course of CD [28, 34, 35].

Therefore, several efforts have been made in this field in order to decrease radiation dose.

Reducing as much as possible the number of enhanced scans and lowering peak kilovoltage (to a maximum of 80 or 100 kV rather than the standard 120 kV) are the simplest strategies to perform [28].

The increased image noise related to the latter approach can be partially compensated by the use of iterative reconstructions that have demonstrated an optimal improvement of the final image quality [20, 34–37].

Also DECT can be of great help thanks to the possibility of iodine recognition and subtraction, with reconstruction of virtual noncontrast (VNC) images from enhanced scans. This allows avoiding the initial unenhanced scan with a consequent spare of radiation delivery [20].

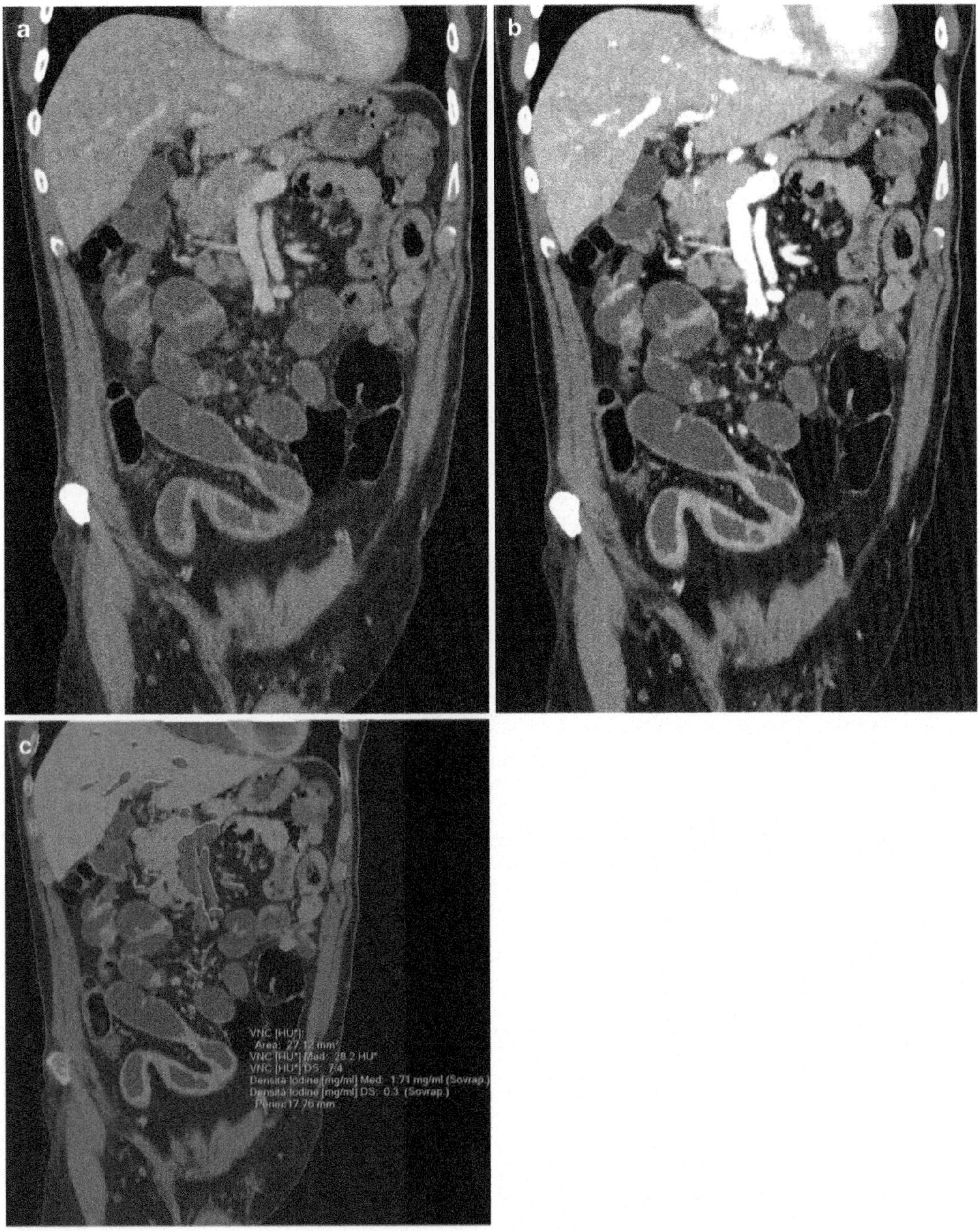

Fig. 2.3 Dual-energy CT-Enterography performed in a CD patient. Conventional venous/enterographic acquisition on coronal plane (**a**) shows wall thickening of the last ileal loop. Virtual monoenergetic image at 50 keV (**b**) and virtual-non-contrast + iodine density (**c**) reconstructions further highlight the mural contrast enhancement. Region of interest (ROI) placed within the mural thickening shows iodine density values

References

1. McNeeley MF, Itani M, Rohrmann CA. Diagnostic fluoroscopy for imaging Crohn's disease. In: Fichera A, Krane M, editors. Crohn's disease. Cham: Springer; 2015.
2. Albert JG, Martiny F, Krummenerl A, Stock K, Lesske J, Göbel CM, Lotterer E, Nietsch HH, Behrmann C, Fleig WE. Diagnosis of small bowel Crohn's disease: a prospective comparison of capsule endoscopy with magnetic resonance imaging and fluoroscopic enteroclysis. Gut. 2005;54(12):1721–7.
3. Lee SS, Kim AY, Yang SK, Chung JW, Kim SY, Park SH, Ha HK. Crohn disease of the small bowel: comparison of CT enterography, MR enterography, and small-bowel follow-through as diagnostic techniques. Radiology. 2009;251(3):751–61.
4. Saibeni S, Rondonotti E, Iozzelli A, Spina L, Tontini GE, Cavallaro F, Ciscato C, de Franchis R, Sardanelli F, Vecchi M. Imaging of the small bowel in Crohn's disease: a review of old and new techniques. World J Gastroenterol. 2007;13(24):3279–87.
5. Haas K, Rubesova E, Bass D. Role of imaging in the evaluation of inflammatory bowel disease: how much is too much? World J Radiol. 2016;8(2):124–31.
6. Cicero G, Mazziotti S. Crohn's disease at radiological imaging: focus on techniques and intestinal tract. Intest Res. 2020; Epub ahead of print
7. Gatta G, Di Grezia G, Di Mizio V, Landolfi C, Mansi L, De Sio I, Rotondo A, Grassi R. Crohn's disease imaging: a review. Gastroenterol Res Pract. 2012;2012:816920.
8. Jaffe TA, Gaca AM, Delaney S, Yoshizumi TT, Toncheva G, Nguyen G, Frush DP. Radiation doses from small-bowel follow-through and abdominopelvic MDCT in Crohn's disease. AJR Am J Roentgenol. 2007;189(5):1015–22.
9. Bungay H. Small bowel imaging in Crohn's disease. Frontline Gastroenterol. 2012;3(1):39–46.
10. Conti CB, Giunta M, Gridavilla D, Conte D, Fraquelli M. Role of bowel ultrasound in the diagnosis and follow-up of patients with Crohn's disease. Ultrasound Med Biol. 2017;43(4):725–34. https://doi.org/10.1016/j.ultrasmedbio.2016.12.014. Epub 2017 Feb 6
11. Gonzalez-Montpetit E, Ripollés T, Martinez-Pérez MJ, Vizuete J, Martín G, Blanc E. Ultrasound findings of Crohn's disease: correlation with MR enterography. Abdom Radiol (NY). 2021;46(1):156–67. https://doi.org/10.1007/s00261-020-02622-3. Epub 2020 Jun 30
12. Bhatnagar G, Von Stempel C, Halligan S, Taylor SA. Utility of MR enterography and ultrasound for the investigation of small bowel Crohn's disease. J Magn Reson Imaging. 2017;45(6):1573–88. https://doi.org/10.1002/jmri.25569. Epub 2016 Dec 9
13. Bollegala N, Griller N, Bannerman H, Habal M, Nguyen GC. Ultrasound vs endoscopy, surgery, or pathology for the diagnosis of small bowel Crohn's disease and its complications. Inflamm Bowel Dis. 2019;25(8):1313–38. https://doi.org/10.1093/ibd/izy392.
14. Carnevale Maffè G, Brunetti L, Formagnana P, Corazza GR. Ultrasonographic findings in Crohn's disease. J Ultrasound. 2014;18(1):37–49. https://doi.org/10.1007/s40477-014-0096-3. PMID: 25767639; PMCID: PMC4353828
15. Kucharzik T, Maaser C. Intestinal ultrasound and management of small bowel Crohn's disease. Ther Adv Gastroenterol. 2018;11:1756284818771367. https://doi.org/10.1177/1756284818771367. PMID: 29881463; PMCID: PMC5987904
16. Lu C, Merrill C, Medellin A, Novak K, Wilson SR. Bowel ultrasound state of the art: grayscale and doppler ultrasound, contrast enhancement, and elastography in Crohn disease. J Ultrasound Med. 2019;38(2):271–88. https://doi.org/10.1002/jum.14920. Epub 2019 Jan 3
17. Coelho R, Ribeiro H, Maconi G. Bowel thickening in Crohn's disease: fibrosis or inflammation? Diagnostic ultrasound imaging tools. Inflamm Bowel Dis. 2017;23(1):23–34. https://doi.org/10.1097/MIB.0000000000000997.
18. Mocci G, Migaleddu V, Cabras F, Sirigu D, Scanu D, Virgilio G, Marzo M. SICUS and CEUS imaging in Crohn's disease: an update. J Ultrasound. 2017;20(1):1–9. https://doi.org/10.1007/s40477-016-0230-5. PMID: 28298939; PMCID: PMC5334271
19. Grażyńska A, Kufel J, Dudek A, Cebula M. Shear wave and strain elastography in Crohn's disease-a systematic review. Diagnostics (Basel). 2021;11(9):1609. https://doi.org/10.3390/diagnostics11091609. PMID: 34573952; PMCID: PMC8468946
20. Sheedy SP, Kolbe AB, Fletcher JG, Fidler JL. Computed tomography enterography. Radiol Clin N Am. 2018;56(5):649–70. https://doi.org/10.1016/j.rcl.2018.04.002. Epub 2018 Jul 11
21. Phillips A. Crohn's disease: CT enterography and CT enteroclysis. In: Hamm B, Ros PR, editors. Abdominal imaging. Berlin, Heidelberg: Springer; 2013. https://doi.org/10.1007/978-3-642-13327-5_221.
22. Gajendran M, Loganathan P, Catinella AP, Hashash JG. A comprehensive review and update on Crohn's disease. Dis Mon. 2018;64(2):20–57. https://doi.org/10.1016/j.disamonth.2017.07.001. Epub 2017 Aug 18
23. Liu W, Liu J, Xiao W, Luo G. A diagnostic accuracy meta-analysis of CT and MRI for the evaluation of small bowel Crohn disease. Acad Radiol. 2017;24(10):1216–25. https://doi.org/10.1016/j.acra.2017.04.013. Epub 2017 Jun 5
24. Gale HI, Sharatz SM, Taphey M, Bradley WF, Nimkin K, Gee MS. Comparison of CT enterography and MR enterography imaging features of active Crohn disease in children and adolescents. Pediatr Radiol. 2017;47(10):1321–8. https://doi.org/10.1007/s00247-017-3876-z. Epub 2017 May 3
25. Sokhandon F, Al-Katib S, Bahoura L, Copelan A, George D, Scola D. Multidetector CT enterography of focal small bowel lesions: a radiological-pathological

correlation. Abdom Radiol (NY). 2017;42(5):1319–41. https://doi.org/10.1007/s00261-016-1015-1.

26. Wong J, Moore H, Roger M, McKee C. CT enterography: mannitol versus VoLumen. J Med Imaging Radiat Oncol. 2016;60(5):593–8. https://doi.org/10.1111/1754-9485.12486. Epub 2016 Jul 28
27. Ilangovan R, Burling D, George A, Gupta A, Marshall M, Taylor SA. CT enterography: review of technique and practical tips. Br J Radiol. 2012;85(1015):876–86. https://doi.org/10.1259/bjr/27973476. Epub 2012 May 2. PMID: 22553291; PMCID: PMC3474054
28. Elsayes KM, Al-Hawary MM, Jagdish J, Ganesh HS, Platt JF. CT enterography: principles, trends, and interpretation of findings. Radiographics. 2010;30(7):1955–70. https://doi.org/10.1148/rg.307105052.
29. Guler E, Unal NG, Hekimsoy I, Kose T, Harman M, Ozutemiz AO, Elmas NZ. Dual-energy CT enterography in evaluation of Crohn's disease: the role of virtual monochromatic images. Jpn J Radiol. 2021;39(4):341–8. https://doi.org/10.1007/s11604-020-01065-6. Epub 2020 Nov 7
30. Lee SM, Kim SH, Ahn SJ, Kang HJ, Kang JH, Han JK. Virtual monoenergetic dual-layer, dual-energy CT enterography: optimization of keV settings and its added value for Crohn's disease. Eur Radiol. 2018;28(6):2525–34. https://doi.org/10.1007/s00330-017-5215-z. Epub 2018 Jan 2
31. D'Angelo T, Cicero G, Mazziotti S, Ascenti G, Albrecht MH, Martin SS, Othman AE, Vogl TJ, Wichmann JL. Dual energy computed tomography virtual monoenergetic imaging: technique and clinical applications. Br J Radiol. 2019;92(1098):20180546. https://doi.org/10.1259/bjr.20180546. Epub 2019 Apr 9. PMID: 30919651; PMCID: PMC6592074
32. De Kock I, Delrue L, Lecluyse C, Hindryckx P, De Vos M, Villeirs G. Feasibility study using iodine quantification on dual-energy CT enterography to distinguish normal small bowel from active inflammatory Crohn's disease. Acta Radiol. 2019;60(6):679–86. https://doi.org/10.1177/0284185118799508. Epub 2018 Sep 5
33. Kim YS, Kim SH, Ryu HS, Han JK. Iodine quantification on spectral detector-based dual-energy CT Enterography: correlation with Crohn's disease activity index and external validation. Korean J Radiol. 2018;19(6):1077–88. https://doi.org/10.3348/kjr.2018.19.6.1077. Epub 2018 Oct 18. PMID: 30386139; PMCID: PMC6201976
34. Park SH, Ye BD, Lee TY, Fletcher JG. Computed tomography and magnetic resonance small bowel enterography: current status and future trends focusing on Crohn's disease. Gastroenterol Clin N Am. 2018;47(3):475–99. https://doi.org/10.1016/j.gtc.2018.04.002. Epub 2018 Jul 7
35. Camera L, Pezzullo F, Acampora A, Liuzzi R, Rispo A, Nardone OM, Luglio G, Bucci L, Castiglione F, Brunetti A. Multi-detector CT enterography in active inflammatory bowel disease: image quality and diagnostic efficacy of a low-radiation high contrast protocol. Clin Imaging. 2019;58:27–33. https://doi.org/10.1016/j.clinimag.2019.06.007. Epub 2019 Jun 14
36. Son JH, Kim SH, Cho EY, Ryu KH. Comparison of diagnostic performance between 1 millisievert CT enterography and half-standard dose CT enterography for evaluating active inflammation in patients with Crohn's disease. Abdom Radiol (NY). 2018;43(7):1558–66. https://doi.org/10.1007/s00261-017-1359-1.
37. Rosenfeld G, Brown J, Vos PM, Leipsic J, Enns R, Bressler B. Prospective comparison of standard- versus low-radiation-dose CT enterography for the quantitative assessment of Crohn disease. AJR Am J Roentgenol. 2018;210(2):W54–62. https://doi.org/10.2214/AJR.17.18296. Epub 2017 Dec 20

3 MR-Enterography: Technique

Silvio Mazziotti, Tommaso D'Angelo,
Giorgio Ascenti, and Giuseppe Cicero

Evaluation of the gut through MRI comprises two different techniques which differ for the administration route of the enteric contrast medium [1].

MR enteroclysis (MREc) foresees the positioning of a nasoenteric tube, through which the intraluminal contrast medium is instilled [2–4].

The positioning of a nasojejunal balloon-tipped catheter can be accomplished through fluoroscopic or MRI guidance [4, 5]. Enteric contrast material can be administered by manual injection with handheld MR-compatible infusion devices or with automated pumps, while the patient is in the MR scanner [4, 5]. The volume and the speed of infusion are crucial for the success of examination.

Depending on the subject, the volume generally varies between 1500 and 3000 ml with an infusion rate ranging from 80 to 200 ml/min. In MREc, the enteric contrast medium can be divided in two phases: (a) A low infusion rate of 80–150 ml/min is used during the first phase, which lasts until terminal ileum begins to distend; (b) In the second phase, the infusion rate increases up to 200 ml/min to achieve reflex atony (note that if the infusion rate increases too fast, retrograde filling can often occur, and this may result in patient's vomiting). A third phase, with an infusion rate decreased again to 80–100 ml/min, can be added in order to guarantee an adequate distention of the proximal jejunum until the acquisition of cross-sectional images is performed [6, 7].

MREc provides optimal distention of the bowel wall and can show detailed luminal information useful to identify early mural changes. However, these advantages have to be counterbalanced by the complexity of the procedure and by the associated patient discomfort [8]. Moreover, an excessive distention of the bowel loops can lead to a worse assessment of the mesenteric structures, which may be compressed by nearby loops.

On the other hand, Magnetic Resonance Enterography (MREg) technique has been developed as a noninvasive alternative to MREc, due to the fact that a significant portion of patients refuses nasojejunal catheter placement. Enterography technique requires the ingestion of a large amount of fluid that fills the stomach and the small bowel in continuity.

Several different ingestion algorithms have been proposed, and they are determined by the bowel transit time of the contrast agent used and by the amount given.

In the case of PEG electrolyte solution, a total amount of 1500–1800 ml is generally administered within 45–60 min prior to scanning. More prolonged timing of administration can lead to colonic filling with impairment of small bowel evaluation.

S. Mazziotti · T. D'Angelo · G. Ascenti
G. Cicero (✉)
Department of Biomedical Sciences and Morphological and Functional Imaging, University Hospital Messina, Messina, Italy
e-mail: smazziotti@unime.it; tdangelo@unime.it; gascenti@unime.it; gcicero@unime.it

S. Mazziotti et al. (eds.), *MR Enterography*, https://doi.org/10.1007/978-3-031-11930-9_3

For an improved assessment of the jejunal loops, a split-acquisition protocol has been proposed, with performance of early scans at 20 min after the start of contrast medium ingestion and delayed acquisitions (approximately at 45 min) for the evaluation of the distal ileum. However, a unique acquisition at 30–40 min is generally adequate [7, 9].

With regard to pediatric patients, the volume of oral contrast is usually adjusted between 600 and 1000 ml according to the weight of the subject [10, 11].

Nausea and vomiting caused by oral contrast ingestion are reduced by antiemetic medication so that examination can be comfortably performed.

Although we aim for a total of 1500–1800 ml, some patients cannot tolerate this volume ingestion, and adequate results may be still achieved with as little as 500–600 ml. For this reason, the examination could proceed even if only a small volume of oral contrast has been ingested.

Due to the poor tolerability of nasojejunal tube insertion or to excessive and uncomfortable bowel distention of MREc, in our practice we recur to enteroclysis only for the evaluation of patients unable or unwilling to ingest oral contrast for enterography [12].

The addition of a rectal water enema provides a better visualization of the terminal ileum and may be considered an adjunct to facilitate the evaluation of colon. However, rectal enema is not routinely performed because the colon is readily accessible at colonoscopy; anterograde filling is also possible and may be more tolerable.

3.1 Enteric Contrast Agents

An adequate intestinal dilation is the main technical requirement for an optimal small bowel imaging, as decades of experience acquired by conventional radiology with barium contrast reveal.

In order to achieve luminal distension, different enteric contrast materials have been investigated for MREg and MREc.

The main purpose is to obtain uniform luminal distention with minimal mucosal absorption and lack of adverse effects or artifacts.

Contrast agents can be classified into three main categories, according to their signal intensity on the pulse sequences used (Fig. 3.1) [2, 5].

Positive contrast agents are characterized by hyperintensity on both T1- and T2-weighted images. They consist of paramagnetic substances, such as gadolinium chelates, ferrous ammonium citrates, manganese chloride, and food products (e.g., blueberry juice, pineapple juice) [13–15].

These contrast agents reduce T1 relaxation time, while T2 relaxation time is usually not affected. However, due to the high water content of their solutions, they are also visualized as hyperintense on T2-weighted images [16]. Positive contrast agents are generally disfavored because their high T1 signal interferes with evaluation of inflammatory mucosal and mural enhancement or of intraluminal lesions after intravenous (i.v.) contrast administration.

Negative contrast agents are characterized by hypointensity at T1- and T2-weightings. They consist of superparamagnetic particles such as iron oxides, perfluorooctyl bromide, and oral magnetic particles [4, 13, 14, 17–20].

Barium sulfate can also be used as a negative agent when administered at high concentrations, but the signal loss is not as great as with the superparamagnetic particles [16]. These contrast agents induce local inhomogeneity in the magnetic field, affecting both T1 and T2 relaxation times. They may improve the conspicuity of wall edema on T2-weighted sequences as well as the detection of extraluminal fluid collections. Due to their intraluminal loss of signal intensity, bowel wall enhancement will be more remarkable on T1-weighted images.

Superparamagnetic particles have a small incidence of side effects (5–15% of cases), which are mainly limited to the gastrointestinal area (suboptimal palatability, nausea, vomiting, and rectal leakage) [5].

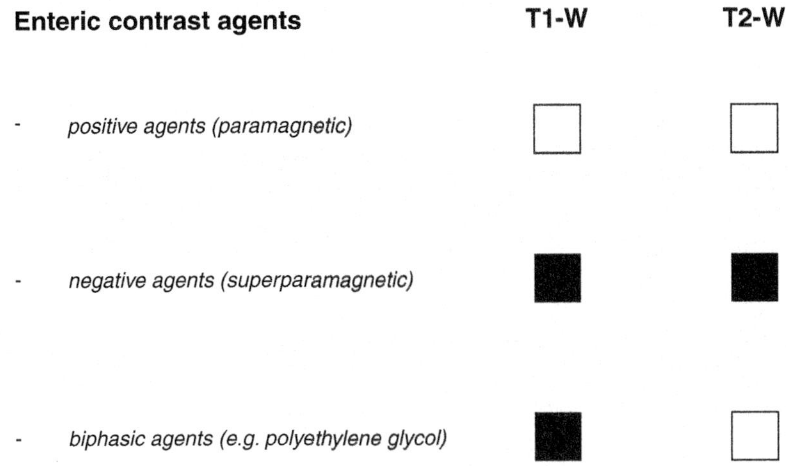

Fig. 3.1 Classification of oral contrast agents. *Positive contrast agents*: high signal intensity on both T1- and T2-w images. *Negative contrast agents*: low signal intensity on T1- and T2-w images. *Biphasic contrast agents*: low signal intensity on T1-w images and high signal intensity on T2-w images

Moreover, if negative contrast agents are not homogeneously distributed throughout the bowel loops, they can produce paradoxical high signal intensity.

The preferences of many authors are currently moved towards the use of *biphasic contrast agents* and in particular those characterized by low signal intensity on T1-weighted images and high signal intensity on T2-weighted images [4]. Following i.v. contrast medium injection, the low signal intensity on T1-weighted images provides a better contrast between bowel lumen and walls, especially in case of hyperenhancement due to inflammation or masses.

On T2-weighted images, these agents allow identification of number and thickness of folds and highlight mural ulcerations.

The biphasic category contains the largest number of available agents, including water, osmotic agents, non-osmotic bulking agents, and polyethylene glycol, each having unique advantages and limitations.

Although water alone is promptly available, better accepted by the patients and cheap, it is rapidly absorbed from distal bowel; therefore, adequate distention may not be obtained.

To obviate this problem, osmotic agents such as mannitol are needed, which however cause osmotic effects in the small bowel like diarrhea, meteorism, and abdominal cramps [21, 22].

Non-osmotic bulk agents such as locus bean gum and methylcellulose have been also evaluated; however, they are not widely available [15, 22–24].

Polyethylene glycol (PEG) solution is a hydrophilic molecule with a low partition coefficient, which determines an inconsistent transmembranous diffusion in the lipid phase. It also has a transversal diameter that does not consent intestinal absorption. Therefore, PEG simulates the properties of water with the advantage of non-absorbability, providing good distention of the entire small bowel. The main problem of PEG solution is that it can lead to rapid bowel transit and a strong urge to evacuate that can interfere with completion of the examination [4, 18, 25–27].

Among the other biphasic agents, manganese and gadolinium chelates must be considered; in fact they are seen as low signal intensity on T2-weighted images and high signal intensity on T1-weighted images when administered at high concentrations [13].

Barium products at lower weight per volume can also act as biphasic contrast agents being well tolerated by patients and providing good intestinal distention [5].

3.2 Patient's Preparation and Positioning

Adequate patient's preparation is mandatory for achieving an optimal intestinal distention.

In order to do so, patients undergoing MREg/MREc should fast for 6–8 h before the start of the exam.

Conversely, bowel cleansing is not required.

Sedation is avoided due to the need of patient's collaboration for assuming enteric contrast agent.

In order to increase signal-to-noise ratio, the patient is imaged using an abdominal phased-array radio-frequency surface coil, in either the supine or prone position (if no stoma is present).

Although MRI is routinely performed in supine position for a better patient's comfort, the preferred scanning position should be prone in order to facilitate elevation and separation of small bowel loops out of the pelvis. Moreover, this position produces a degree of abdominal compression, reducing the number of sections required for each coronal acquisition, which in turn reduces the length of breath-hold required, resulting in improved patient compliance (Fig. 3.2). It has been also shown that it improves small bowel distention (Fig. 3.3) and that this position is safer if the patient should vomit. There are no specific contraindications to MREg/MREc, except those typical for MR (e.g., metal implants in delicate positions, aneurysm clips, shrapnel injuries, pacemakers, internal defibrillators). A relative contraindication to MR examination could be considered the inability to receive gadolinium-based contrast medium (patients with a low glomerular filtration rate, who are at risk for nephrogenic systemic fibrosis or pregnant patients). However, it should be noted that in these cases the examination could be performed without i.v. contrast material, allowing one to obtain useful findings in CD.

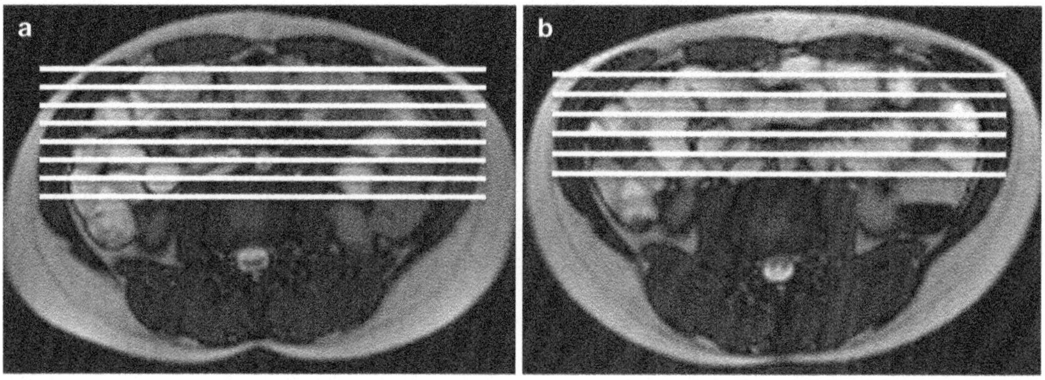

Fig. 3.2 Axial scout images obtained in supine (**a**) and prone position (**b**). Abdominal compression consequent to prone position (**b**) allows to reduce the anteroposterior diameter and to acquire a reduced number of coronal slices in respect to supine position (**a**)

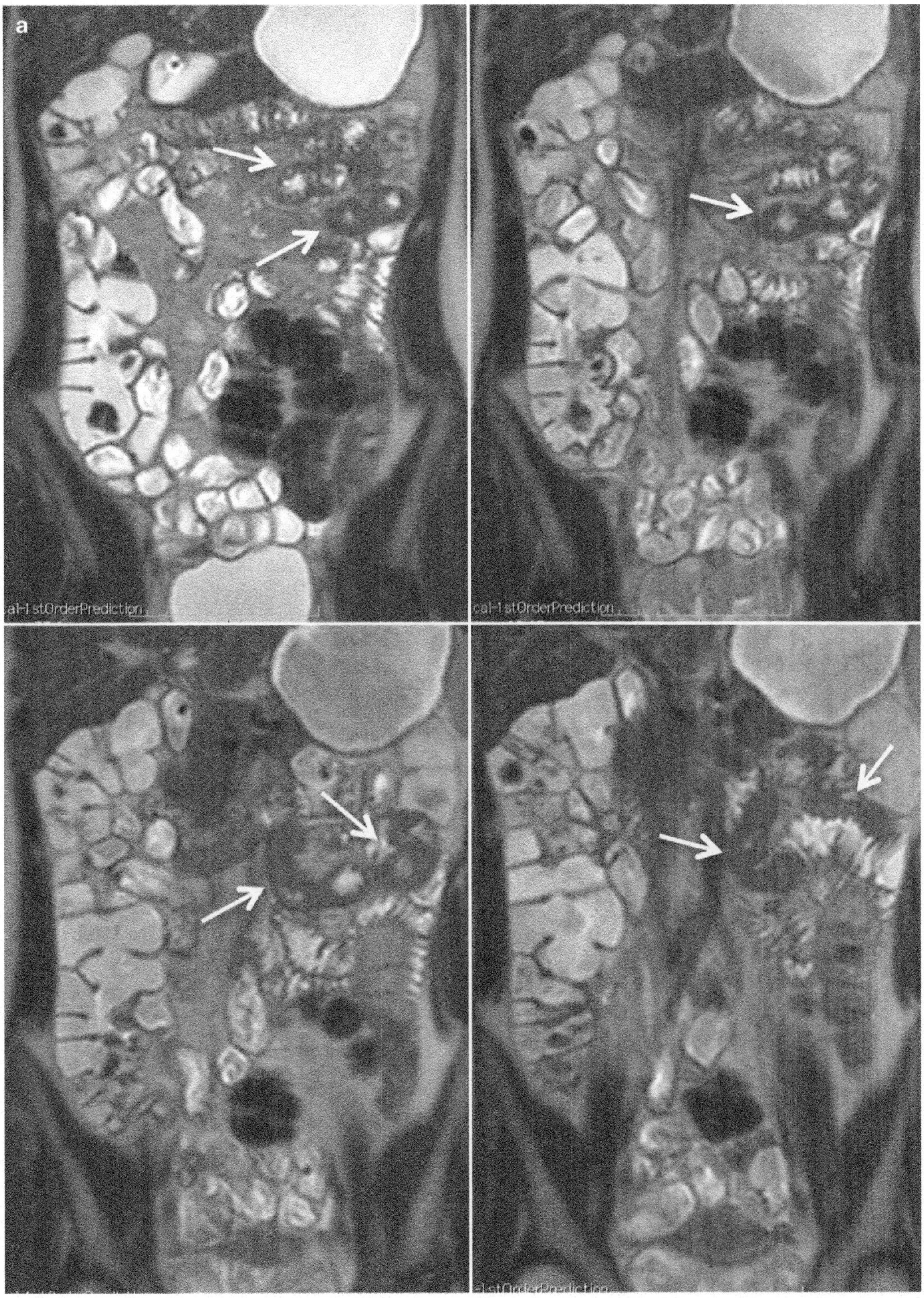

Fig. 3.3 Coronal T2-w HASTE images obtained in the same patient in supine (**a**) and prone position (**b**). In supine position (**a**) the poor distention of jejunal and proximal ileal segments can lead to difficulties of interpretation, as the result of apparent thickening of the bowel wall (arrows). Prone imaging (**b**) results in significantly higher small bowel distention and better bowel loops separation, excluding any pathological wall thickening

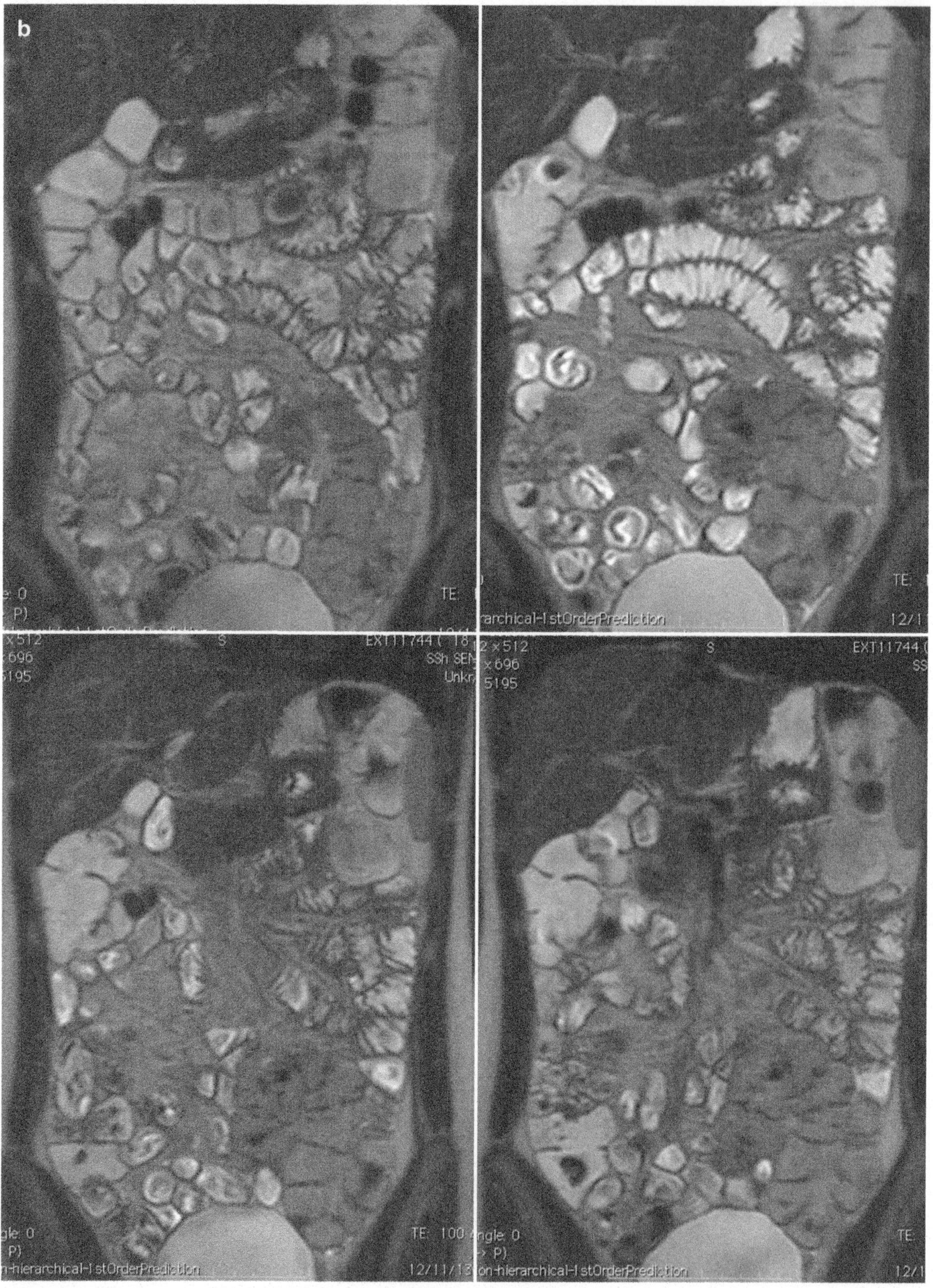

Fig. 3.3 (continued)

3.3 Protocols and Sequences

In the past, motion artifacts and poor contrast resolution precluded the use of MR for bowel imaging. Technological advances, including the use of breath-hold sequences, improved coils, fat suppression, and gadolinium, have extended the role of MRI in the evaluation of gastrointestinal tract [5]. Since then, several different pulse sequences to evaluate the small bowel have been advocated by different authors.

Although protocols may vary due to differences in equipment and personal expertise, three main type sequences are generally acquired on coronal and axial planes:

- Half-Fourier acquisition single-shot turbo spin echo with and without fat suppression.
- Balanced steady-state free precession.
- Pre- and post-contrast T1-weighted ultrafast gradient echo [28–30].

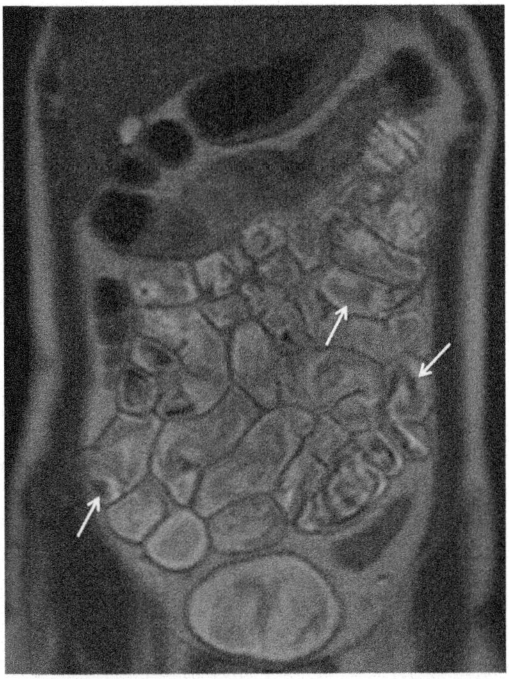

Fig. 3.4 Coronal T2-w HASTE image. Normal small bowel. Intraluminal flow-voids, secondary to bowel peristalsis, are seen (arrows)

3.3.1 T2-Weighted Scans

3.3.1.1 Half-Fourier Single-Shot Turbo Spin Echo (HASTE)

HASTE sequence is an adaptation of Rapid acquisition with relaxation enhancement (RARE) that reduces total acquisition time by acquiring only half of k-space. Different manufacturers call these sequences *single-shot fast spin echo (SSFSE)*. These sequences provide heavily T2-weighted images in less than 1 s with high contrast between the lumen and the bowel wall. As these sequences are highly resistant to magnetic susceptibility or chemical shift artifacts, the wall thickness may be accurately evaluated [4, 29, 30].

Moreover, sinus tract, fistulas, and fluid collections are well visualized. HASTE sequences are susceptible to motion artifacts of fluids within the visceral lumen (propulsive intestinal movements), which may result in intraluminal low-signal-intensity artifacts and appearance of "pseudo-lesions" (Fig. 3.4) [4]. Due to k-space filtering effects, visualization of the mesenteric structures such as mesenteric vessels and lymph nodes is impaired. The optional use of fat saturation allows differentiation between submucosal fat and edema which appear both bright on T2-weighted images. Fat saturation also highlights bowel edema and improves contrast between the intestinal wall and the surrounding fat tissue. Fat-suppressed HASTE sequences are particularly useful when i.v. contrast material administration is contraindicated to assess parietal inflammatory infiltration and to reveal inflammation of the peritoneal fat tissue [4].

3.3.1.2 Single-Shot Thick-Slab RARE

RARE with an ultralong echo train length scans are heavily T2-weighted sequences that are usually employed during cholangiopancreatography and MR pyelography.

Therefore, these images provide selective static fluid visualization with complete background suppression, representing an ideal tool to examine static fluids in the human body [4].

If included in the MRE protocol, these scans are capable in demonstrating the fluid filling of the whole intestinal loops in a single image (Fig. 3.5).

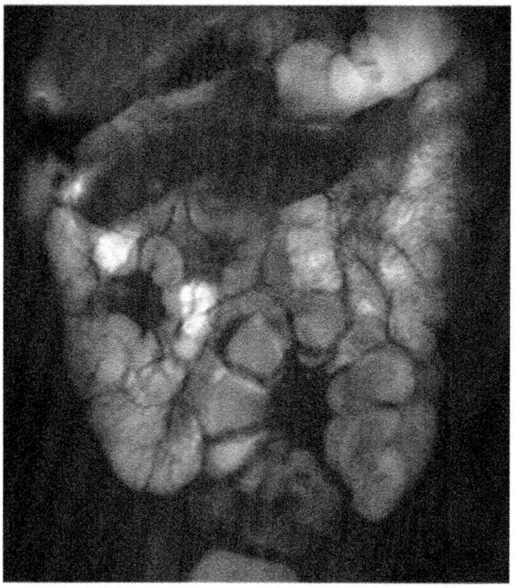

Fig. 3.5 Coronal thick-slab T2-w RARE image (50-mm thickness) obtained with ultralong echo train length showing intestinal distention in a single image

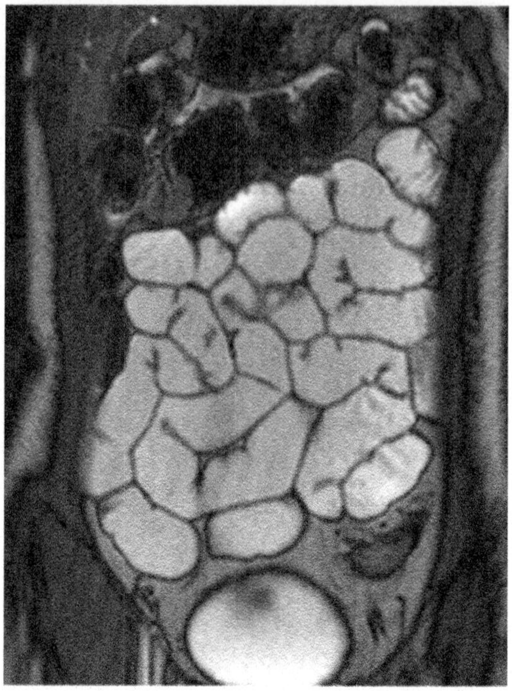

Fig. 3.6 Coronal True-FISP image. Normal small bowel with uniform intraluminal signal and high contrast between bowel wall and lumen. The "*black-boundary*" artifact should not be confused with wall thickening

3.3.2 Balanced Steady-State Free Precession (Balanced SSFP)

Another sequence promoted for imaging of the small bowel is the balanced SSFP, also known as *fast imaging with steady-state acquisition* (*FIESTA*), *balanced fast field echo* (*balanced FFE*), and *true fast imaging with steady-state precession* (*True-FISP*).

This sequence consists of a balanced gradient-echo sequence where image contrast is dependent on T2/T1 ratio for each tissue. The ratio of T2/T1 contrast essentially reflects the T2 differences in tissues as repetition time and echo time are so short that T1 is almost constant: tissues with a T2 value approaching their T1 value appear brightest. Moreover, these sequences eliminate phase shift caused by motion, and thus both fluid and blood appear bright. Due to the extremely short acquisition time, with each image acquired in a few hundred milliseconds, they are relatively insensitive to motion artifacts, providing uniform intraluminal signal and leading to a high contrast between the bowel wall, lumen, and mesentery. In particular, mesenteric vessels and adenopathies are better visualized on these sequences than on the HASTE sequences. A limitation of this sequence is represented by a black-boundary artifact at the interface of the bowel wall and mesenteric fat that may mask small lesions and impede a correct assessment of bowel wall thickening due to a possible overestimation (Fig. 3.6) [22, 28, 29]. This deletion of signal at boundaries is caused by opposed phases of protons in water and fat within a voxel at certain echo times (i.e., when echo times of around 2.3 ms are used at 1.5 T). Fat suppression may help in reducing the effects of the black-boundary artifact. Susceptibility artifact occurs with the presence of intraluminal gas or ferromagnetic material, leading to image distortion. Off-resonance artifacts occur in the presence of a nonuniform magnetic field, resulting in banding artifact at the periphery of the image. Applying built-in shimming procedures during examination can reduce them.

3.3.3 Pre- and Post-Contrast T1-Weighted

3.3.3.1 Ultrafast Gradient Echo

For T1 weighting, gradient-echo sequences are usually performed, as these permit short repetition times, resulting in a very short acquisition time. These can be acquired as either two-dimensional (2D) (*fast spoiled GRASS, fast SPGR*; *turbo fast low angle shot, turbo FLASH*; *turbo field echo, TFE*) or three-dimensional (3D) (*fast acquisition with multiphase EFGRE 3D, FAME*; *volumetric interpolated breath-hold examination, VIBE*; *T1 high-resolution isotropic volume examination, THRIVE*) sequences (Figs. 3.7 and 3.8a).

With a 3D sequence, a radio-frequency pulse excites a thick volume of tissue rather than a thin 2D section. If a large area is covered with 2D acquisitions, separate breath-holds will be required, which can lead to respiratory misregistration. 3D gradient-echo sequences allow thinner collimation. The volumetric data can be reconstructed in any plane, providing increased through-plane and in-plane spatial resolution as well as offering higher signal-to-noise ratio com-

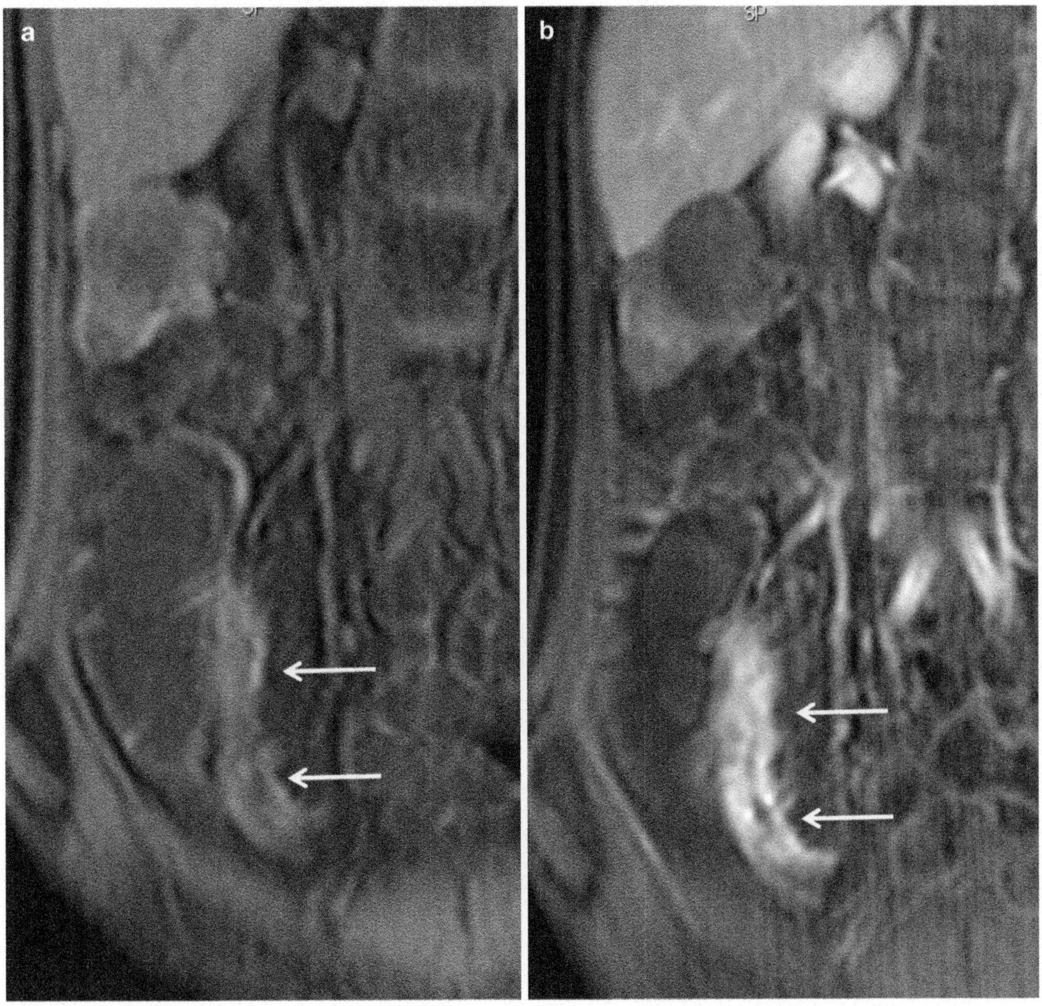

Fig. 3.7 Coronal fat-suppressed 2D GE T1-weigthed image obtained before (**a**) and after i.v. contrast medium administration (**b**) in a patient with active Crohn's disease of terminal ileum (arrows)

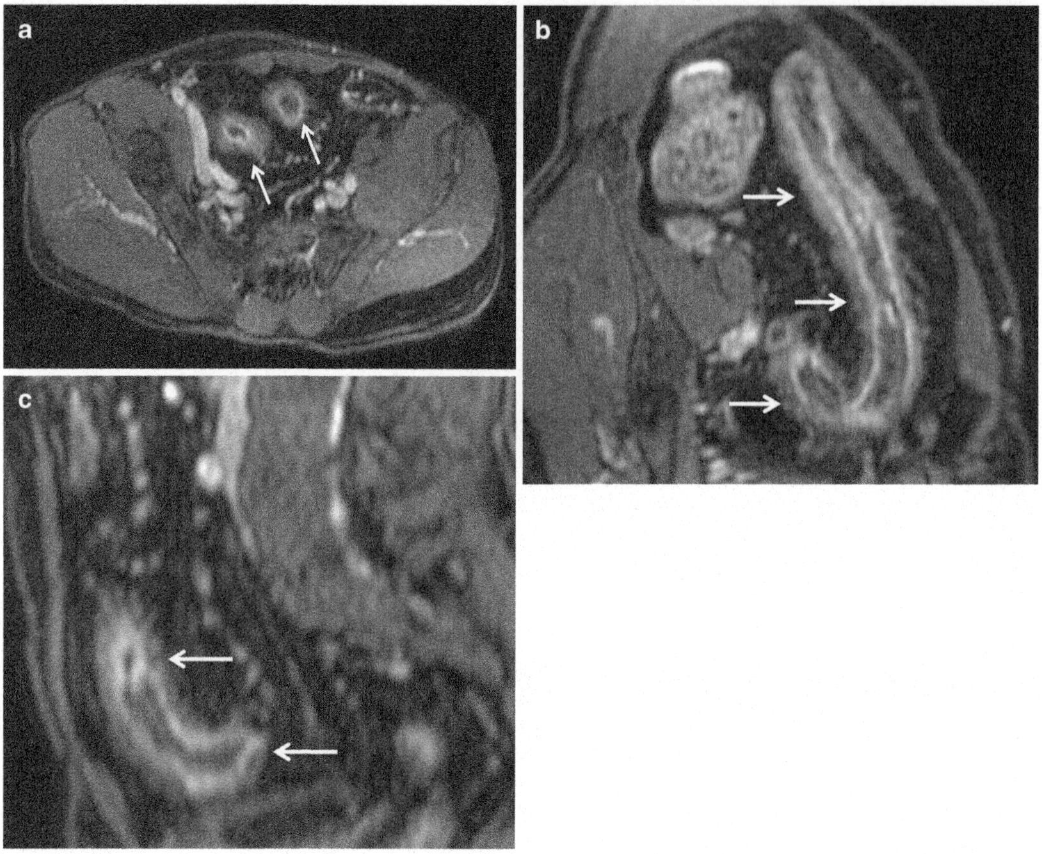

Fig. 3.8 Axial fat-suppressed 3D GE T1-weighted image obtained after i.v. contrast material injection (**a**) in a patient with active inflammation of an ileal loop (arrows). Oblique-coronal (**b**) and sagittal (**c**) reformatted images well depict the length of the pathological ileal segment

pared with 2D sequences (Fig. 3.8b, c). However, because of the relatively greater length of the volumetric acquisition, 3D is more sensible to motion, causing blurring (Figs. 3.9 and 3.10).

Fat saturation can be used to increase contrast resolution facilitating the visualization of the bowel wall, which has high signal intensity. T1-weighted images are routinely obtained before and after i.v. injection of gadolinium.

After i.v. contrast administration, fat saturation also increases the conspicuity of the normal bowel wall allowing the characterization of a lesion by assessment of its enhancement pattern [2, 4, 28–30].

A limitation of gradient-echo sequences in general is the increased sensitivity to magnetic susceptibility effects.

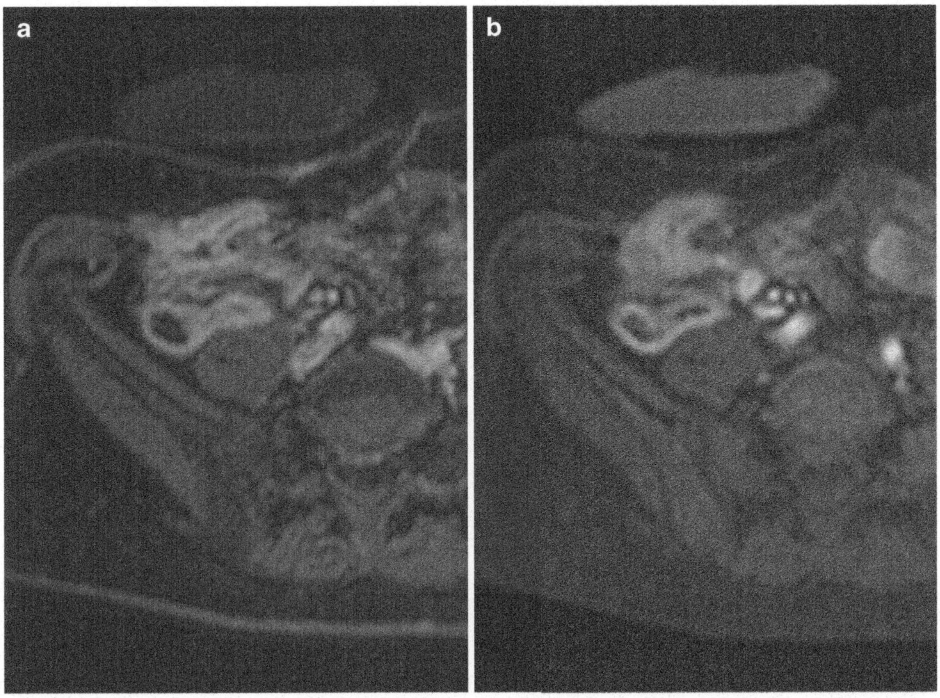

Fig. 3.9 Comparison of 2D and 3D GE T1-weighted sequences in a patient with ileo-colectomy and right ileo-cutaneous stoma. Axial 2D GE T1-weighted image (**a**), and axial 3D GE T1-weighted image (**b**) with fat suppression after i.v. contrast material administration show active inflammation of neoterminal ileum. Note the slight increased blurring of the 3D GE sequence (**b**)

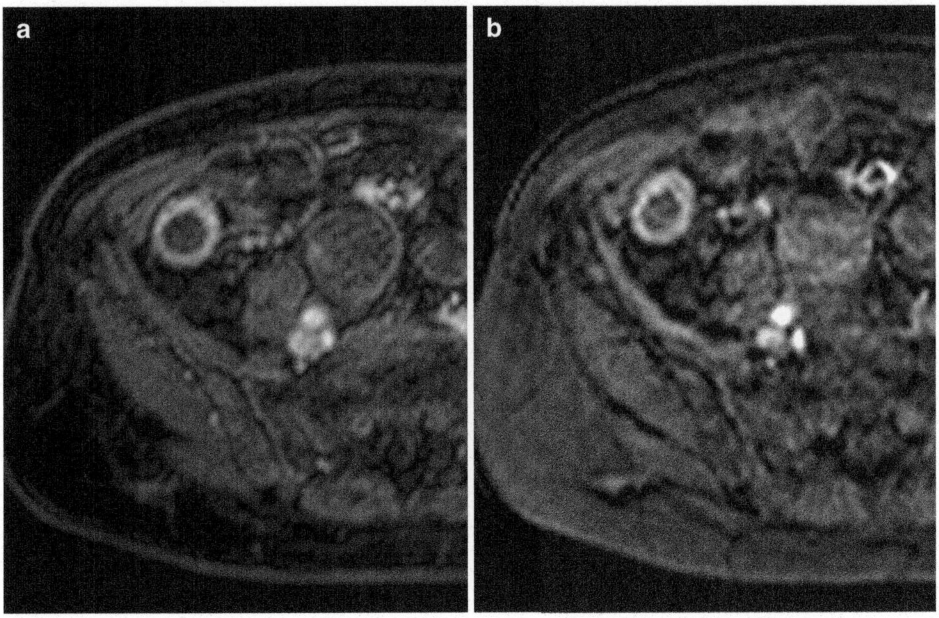

Fig. 3.10 Comparison of 2D and 3D GE T1-weighted sequences in a patient with active Crohn's disease of terminal ileum. Axial 2D GE T1-weighted image (**a**), and axial 3D GE T1-weighted image (**b**) with fat suppression after i.v. contrast material administration show active inflammation of terminal ileum. Note the slight increased blurring of the 3D GE sequence (**b**)

3.4 Spasmolytics

One of the main limitations during the MR examination of the small bowel is represented by the peristaltic motion artifacts, which are more evident in T1-weighted fast-gradient-echo sequences performed after i.v. administration of contrast material (particularly with the motion-sensitive T1-weighted 3D gradient-echo sequences).

These artifacts can potentially hide relevant findings or determine false-positive lesions, such as the intraluminal flow artifacts, previously described in HASTE sequences. Bowel peristalsis can be reduced by i.v. administration of a spasmolytic agent. This usually consists of 20 mg i.v. hyoscine-N-butylbromide (Buscopan, Boehringer Ingelheim, Germany), or when contraindicated or not tolerated (e.g., history of cardiac arrhythmia, narrow angle glaucoma, or prostatism), 1 mg i.v. glucagon is given as alternative (unless patients have a known hypersensitivity to glucagon or a history of pheochromocytoma).

Intravenous (i.v.) administration of spasmolytic agents is usually performed immediately before the acquisition of contrast-enhanced T1-weighted sequences, although some centers reported the use of a double separated injection, one prior the beginning of the examination and a second immediately prior to i.v. contrast administration.

It should be considered that using spasmolytics in routine MRE determines an additional amount of time, due to the patient removal from MR scanner, management of eventual side effects (if they occur), and examination restart.

Moreover, the administration of these agents might be not strictly necessary [31]. In fact it has to be said that the beneficial effects of spasmolytics in MR imaging are more evident in healthy bowel rather than diseased one. This is due to the fact that both fibrosis and bowel wall inflammation determine a decreased bowel motility, thus a less susceptibility to motion artifacts.

Finally, the built-in redundancy of MR examination, which allows evaluation of each bowel segment at multiple time points, can demonstrate whether a questionable segment is affected by disease or if it is just underdistended at the time of image acquisition, further reducing the usefulness of spasmolytic agent (Fig. 3.11).

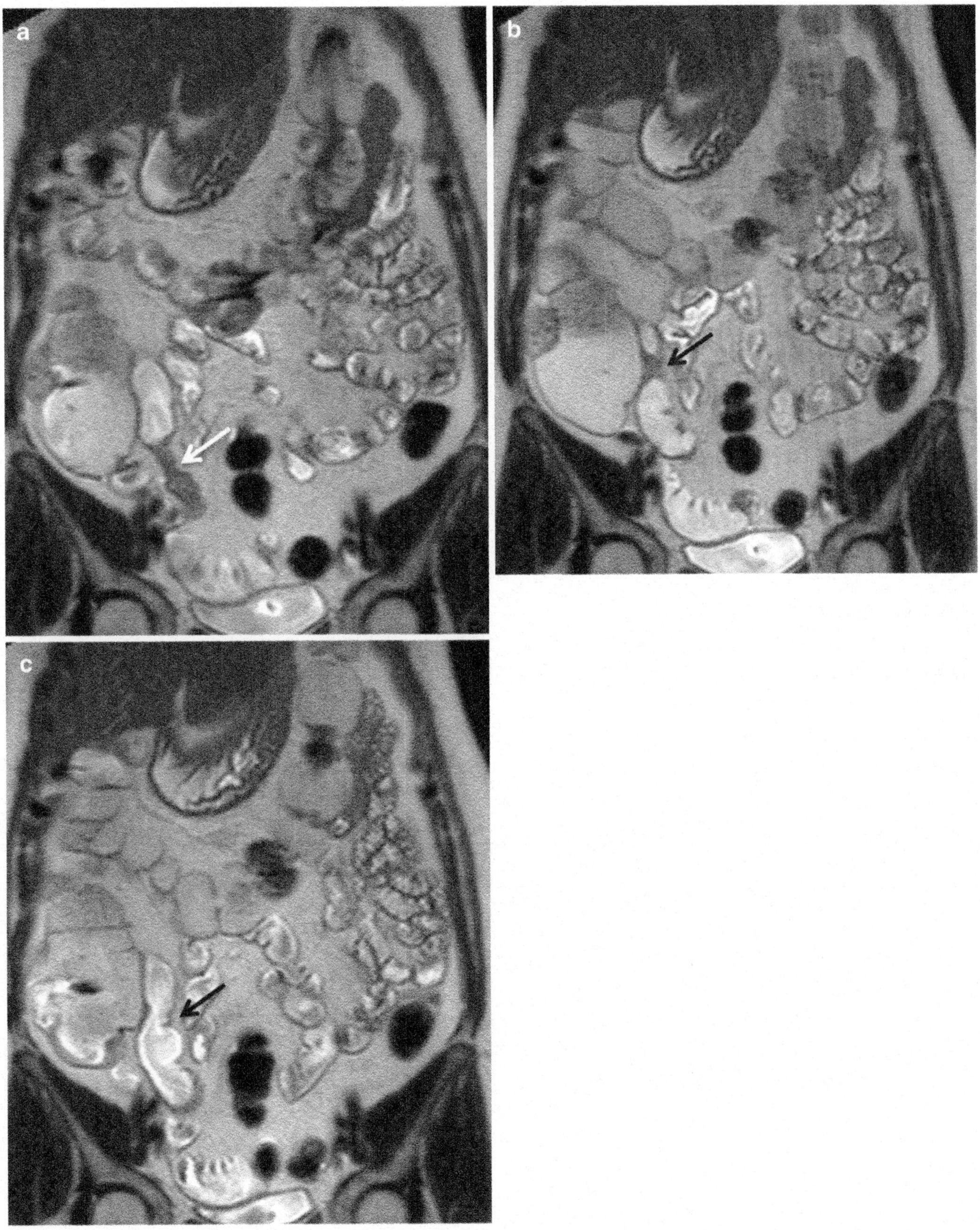

Fig. 3.11 Coronal T2-w HASTE images sequentially acquired at multiple time points (**a, b, c**). In the first image (**a**) a luminal narrowing of the terminal ileum might simulate a fibrotic stricture (arrow); subsequent MR acquisitions (**b, c**) show the distension of the ileal segment, allowing the correct diagnosis of a functional bowel spasm

3.5 Intravenous Contrast Agent

Intravenous (i.v.) administration of gadolinium chelates is very important to assess parietal enhancement of inflamed bowel or lymphadenopathy, and when it occurs, it allows to better delineate sinus tracts and fistulas. T1-weighted images are routinely obtained before and 70 s after i.v. injection of gadolinium at a dosage of 0.2 mmol/kg and a flow rate of 3.5 ml/s, which is followed by a bolus of 10–20 ml isotonic saline at 3.5 ml/s flow rate. Although pump injection of contrast material is optimal, adequate results may be achieved also with hand injection.

Fast-ultrafast gradient-echo sequences performed with both 2D and 3D acquisition are usually chosen, preferably implementing fat saturation to increase contrast resolution and to allow a better assessment of bowel enhancement.

To have a more sensitive visualization of progressive transmural bowel wall enhancement pattern, multiphasic post-contrast acquisition may be performed during the arterial (30 s), portal venous (60–75 s), and interstitial/delayed (3/10 min) phases, using bolus triggering once contrast reaches the descending aorta.

In particular, the fibrotic evolution can be established on the basis of enhancement gain between enterographic and late acquisition, respectively performed 70 s and 7 min after Gd injection [32, 33].

However, it has to be said that gadolinium chelates should not be given in patients with renal impairment, due to the potential long-term risk of nephrogenic systemic fibrosis. Moreover, caution should be taken in case of pregnant patients since gadolinium should be injected only if crucial for the life of the patients or the fetus [34].

3.6 Complementary MR Techniques

Beyond the standard sequences, modified protocols and newer sequences are promising to further improve the diagnostic utility of MR imaging of the small bowel. Assessing small bowel motility alterations and grading the disease activity are the latest goals to reach; for this reason, there is an increasing interest to obtain additional functional information and to calculate quantitative parameters related to histopathological severity of acute inflammation, which besides correlating with clinical findings may predict the clinical course, allowing a rational use of immunosuppressive therapies.

3.6.1 MR Fluoroscopy

As it was previously mentioned, MR fluoroscopy is routinely used during MREc to evaluate oral contrast agent progression in the small bowel [35].

MR fluoroscopy can be performed with a 2D single-shot T2-weighted turbo spin-echo sequence, the same routinely employed in MR cholangiopancreatography and MR pyelography.

Ultralong echo trains in the heavily T2-weighted images provide selective static fluid visualization with complete background suppression. Fat suppression further reduces the remaining signal intensity of the fat so that exclusively fluid is displayed with high signal intensity. Moreover, the use of this sequence in the MR evaluation of the small bowel offers the further advantage of a panoramic view, similar to SBFT, with which the gastroenterologists are usually more confident.

We further developed this technique obtaining MR-fluoroscopy imaging of the small bowel also during routine MREg diagnostic protocol.

This revised technique presumes that the patient, in prone position inside MR scanner, can assume the oral contrast through a drinking straw for the whole duration of fluoroscopic acquisition (Fig. 3.12).

MR-fluoroscopy images are oriented in the coronal plane with a section thickness of 100–180 mm to include the entire intestine and the stomach. Initially, one image is acquired before the oral administration of PEG solution, and subsequent images are obtained every 1–2 min.

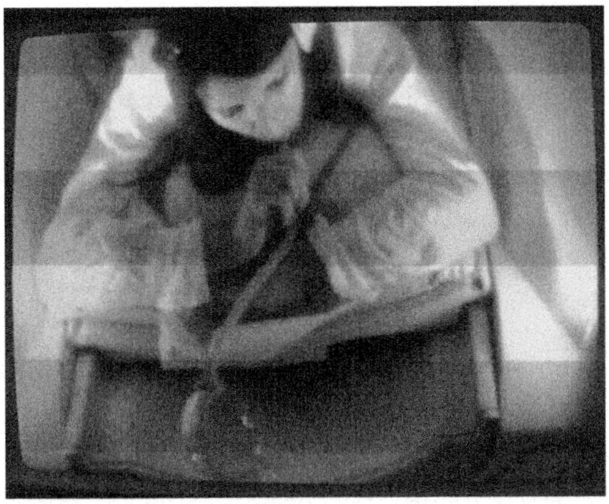

Fig. 3.12 Picture showing the patient, in prone position inside the MR scanner, assuming oral contrast agent from a drinking straw during MREg

This technique provides real-time monitoring and documentation of gastroduodenal and small bowel filling also allowing information about the peristalsis efficiency, thus implementing MREg to better depict proximal jejunal tract (Fig. 3.13) [36].

MR-fluoroscopy images can also be reviewed in a cine-loop format to obtain functional information concerning bowel obstruction.

On MR-fluoroscopy images, fluid-filled bowel loops are not the only structures displayed, but other abdominal static fluid, such as the biliary system, pancreatic duct, urinary tract, as well as ascites, are also depicted.

A large amount of ascites may obscure small bowel loops (Fig. 3.14).

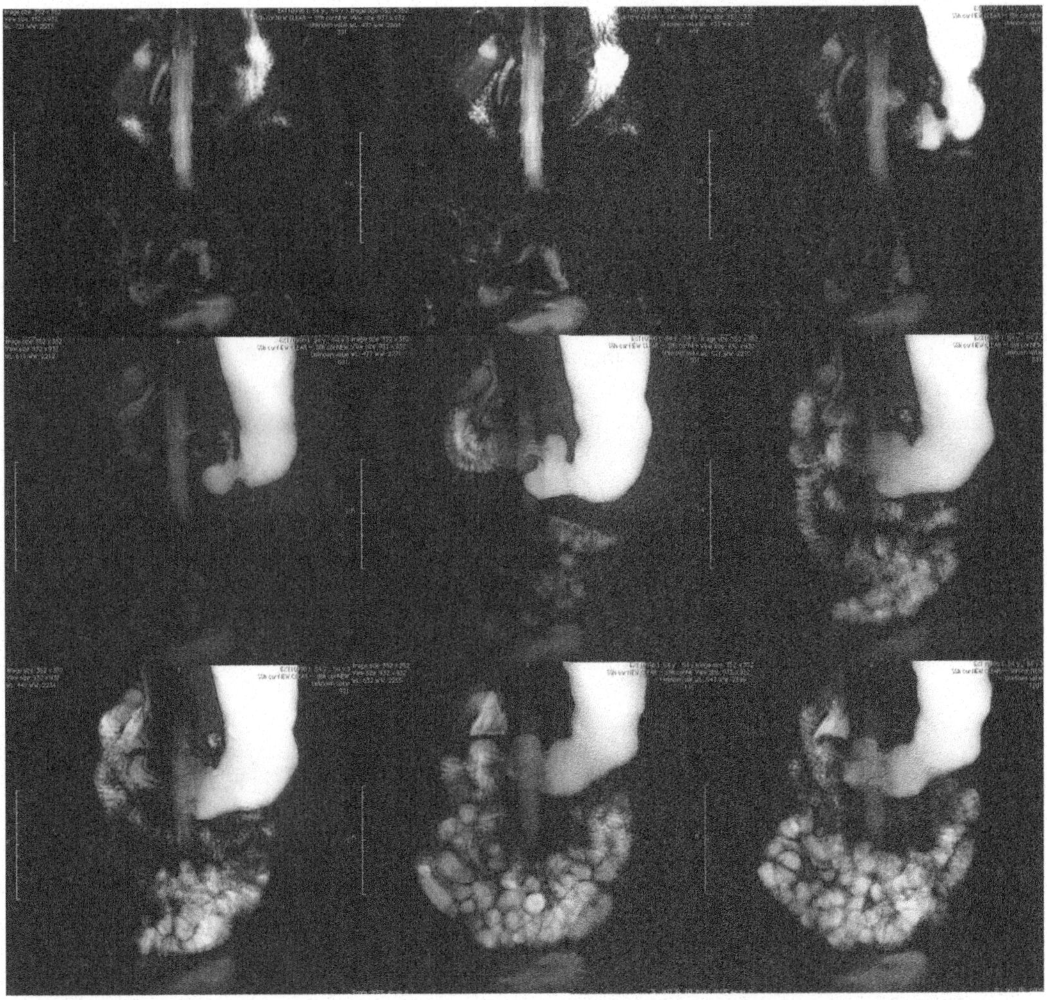

Fig. 3.13 Functional information provided at MR fluoroscopy with progressive filling of the normal small bowel

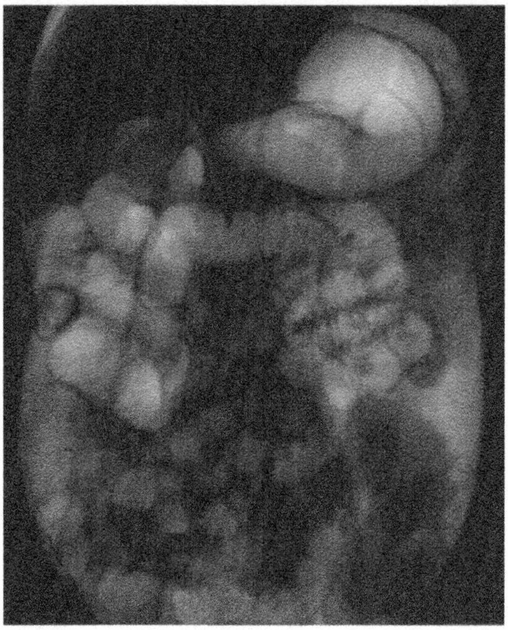

Fig. 3.14 MR-fluoroscopy image in a patient with ascites. The superimposition of hyperintense peritoneal fluid reduces the visibility of bowel loops

3.6.2 Cine MR

In addition to morphological evaluation, MR of the small bowel allows the visualization and analysis of ileal motility with ultrafast imaging techniques (*cine MR*) [37]. In disease states and particularly in CD, inflammatory infiltration of the bowel wall should adversely affect motility. Moreover, additional factors over and above acute inflammation may affect motility, including drug regimens, disease duration, location, coexistence of colitis, previous surgery, and mural fibrosis. Balanced SSFP is an ultrafast imaging sequence providing motion-free images with T2-like contrast, which has been previously used in cardiac imaging.

With technical development, a SSFP sequence allows high temporal resolution imaging of the whole length and width of the abdomen in a coronal slice with subsecond repetition times during a single breath-hold [28]. The resulting cine sequence is suitable to monitor the small bowel motility function. Coronal imaging in prone position is preferably used to separate the bowel loops and also to reduce the displacement of the intestinal structures out of the imaging section. Several cine sequences of 20 frames are performed using a 1-cm-thick coronal two-dimensional balanced SSFP. This sequence has a temporal resolution of 0.5 s indicating that in every 500 ms a new image on the same plane is acquired (Fig. 3.15). Depending on the size of patient, this cine sequence could be repeated stepwise in 15–25 series over the entire abdomen from anterior to posterior without a gap in between.

The cine-MR sequences can be performed repetitively anytime, with total examination time only slightly prolonged for each patient.

The evaluation of the cine sequences can be simply limited to qualitative criteria without measuring the peristaltic frequency; the motility changes are identified as zones of abnormal peristalsis in comparison with surrounding bowel segments.

Recently, software methods have been developed to automatically assess small bowel contractions as well as quantitatively analyze the motility [38, 39]. Such software can quantify changes in luminal diameter over a time series and ultimately may better characterize abnormal patterns of peristalsis in diseased bowel. Post-processing software has been elaborated to facilitate these investigations by reducing the time needed for evaluation of small bowel motility with high reproducibility. Cine-MR sequences are becoming routine in MREg/MREc protocols. Motility, considered as a functional parameter, has been shown to significantly increase the lesion detection rate for CD-related pathological findings in comparison with only morphological MR assessment. Evaluation of small bowel motility could provide additional information about disease status or severity. However, it has to be said that cine MR cannot be used alone as a diagnostic tool. This technique is valuable for showing motility alterations, while the specific CD findings (e.g., wall thickness, stenosis, layering, or contrast enhancement) can only be detected on the static images.

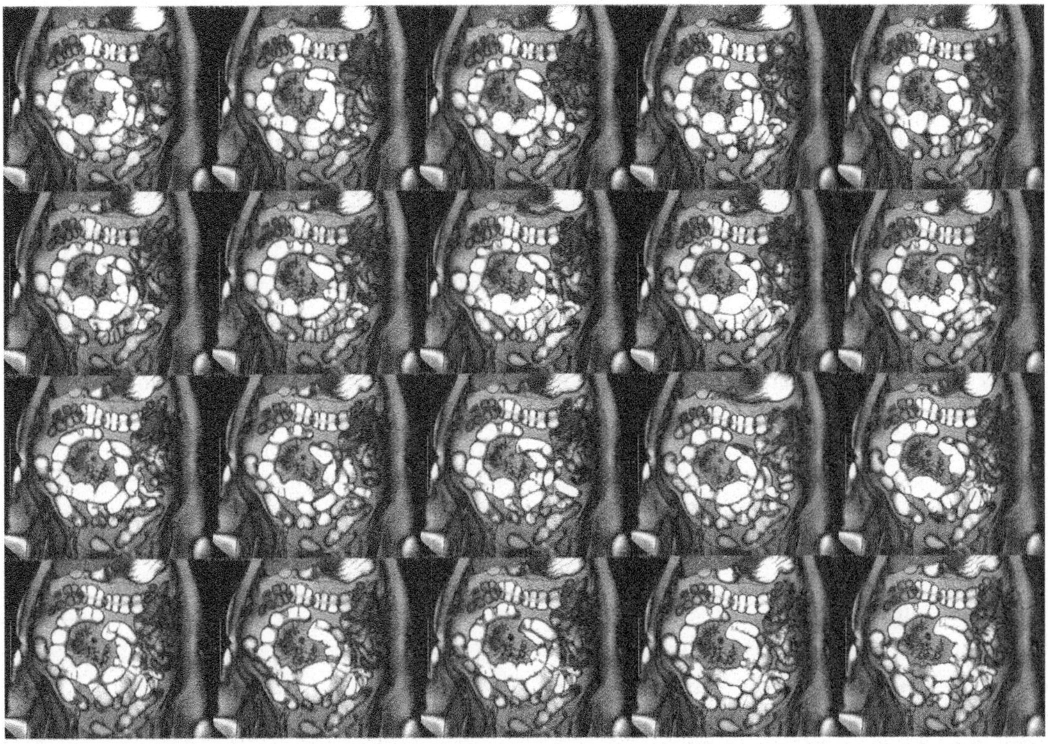

Fig. 3.15 Cine-MR sequences acquired along the same coronal plane which show functional information about small bowel motility

3.6.3 Diffusion-Weighted Imaging

Diffusion-weighted imaging (DWI) is a noninvasive method deriving its image contrast from differences in the motion of water molecules between tissues, and it is sensitive to water molecules exhibiting restricted diffusion.

DWI is increasingly becoming a standard sequence in body MR protocol [29]. In particular, recent hardware and technical developments such as rapid echo planar imaging (EPI), high gradient amplitudes, multichannel coils, and parallel imaging have made DWI a feasible sequence for abdominal imaging. Apparent diffusion coefficient (ADC) can be calculated as a quantitative measure of diffusion [40, 41].

DWI is routinely obtained using a maximum *b value* of 800–1000 s/mm^2 on axial and/or coronal planes. These parameters are used in most of clinical protocols, representing the best compromise between signal-to-noise ratio (SNR) and sensitivity in detecting lesions.

Owing to the presence of the inflammatory infiltrate, intestinal CD lesions are characterized by bright signal in the DW images and lower ADC values compared to normal segments.

However, caution must be taken since similar findings can be detected in case of malignancies, due to the high cellularity of solid tumors and the increased density of cell membranes.

Furthermore, although a number of works have investigated the role of ADC in differentiating active inflammation from fibrosis, a precise cut-off value has not been unanimously identified so far.

3.7 Technical Novelties and Future Perspectives

MRE in the assessment of bowel involvement in CD patients takes benefit technically from the high contrast resolution of soft tissues, the comprehensive evaluation of the abdominal cavity and, from the patient point of view, the lack of invasiveness and absence of ionizing radiations.

However, despite the undoubted advantages, several challenges remain to be faced.

First, the scan protocol is time consuming which results in lower scanner availability and patient comfort, especially for the claustrophobic and the pediatric ones.

In this context, few works have evaluated the possible role of abbreviated scan protocols relying solely on unenhanced acquisitions. These works have shown encouraging results in terms of agreement between readers comparing the abbreviated and the full protocols [42, 43].

Nonetheless, the use of 3 T scanner surely decreases the acquisition time (at least of the 30%), also resulting in less susceptibility to motion artifacts [44].

Second, differentiation of active inflammation from fibrosis is difficult to assess on MRE, since the two phenomena often coexist. However, the achievement of this information could substantially reduce the resort to endoscopy and avoid unnecessary surgical operations.

Perfusion imaging analyzes the time-dependent changes of signal intensity after gadolinium chelate administration by dynamic contrast-enhanced MRI (DCE-MRI) that reflects the status of tissue microcirculation [16, 28, 45].

Using this technique, it is possible to evaluate the enhancement as a function of time and to calculate semiquantitative and quantitative parameters for evaluating perfusion.

In particular, semiquantitative methods refer to the use of parameters that describe the enhancement curve (i.e., initial enhancement, time to peak, slope of the enhancement curve).

On the other hand, quantitative methods rely on pharmacokinetic models to interpret contrast enhancement data in terms of parameters related to the underlying vascular anatomy and physiology [46–48].

Up to now, MR-perfusion in CD patients is a promising technique not only for differentiating affected from unaffected intestinal loops but also for distinguishing active from inactive disease. However, inhomogeneity in literature due to differences among vendors, technical aspects, and evaluated parameters still requires larger validations [49–51].

Magnetization transfer (MT) in a novel MRI sequence that exploits transfer of saturation from immobile protons bound to macromolecules, such as collagen, to mobile protons of free water, allowing image contrast that permits to distinguish the formers from the latter [52–54].

Therefore, MT promises to recognize presence of fibrosis within small bowel strictures.

Initial results derive from rat models and so far only few works have reported their clinical performance on IBD patients [54–56].

Although further confirmations are necessary, MT seems capable not only in identifying fibrosis but also in distinguishing different degrees even when active inflammation is present [57, 58].

Another fascinating future perspective is represented by Magnetic Resonance Spectroscopy (MRS).

MRS is capable in evaluating molecular composition of body tissue and it is routinely performed for assessment of brain, breast, and prostate malignancies [10].

Intestinal mucosa alterations of IBD patients due to inflammation are reflected by synthesis of metabolites, which are then excreted with urine or feces [10].

Therefore, it can be presumed that in the next future MRS could allow a further characterization of the disease stages and perhaps a differentiation between UC and CD just relying on the presence of specific molecules within the intestinal wall [59].

Finally, artificial intelligence (AI), also known as radiomics, is drawing attention within different radiological fields and MRE is not an exception [52, 60].

This method exploits machine learning algorithms in order to extract quantitative, and not only visual, information from raw data.

Initial experiences focused on semiautomatic measurement of wall thickness or motility quantification for identification of areas of active inflammation [61, 62].

On the other hand, texture analysis, through assessment of signal intensity per voxel, has demonstrated encouraging initial results about disease activity assessment and differentiation of inflammation from fibrosis [63, 64].

References

1. Cicero G, Mazziotti S. Crohn's disease at radiological imaging: focus on techniques and intestinal tract. Intest Res. 2021;19(4):365–78. https://doi.org/10.5217/ir.2020.00097. Epub 2020 Nov 25. PMID: 33232590; PMCID: PMC8566824
2. Fidler JL, Guimaraes L, Einstein DM, et al. MR imaging of the small bowel. Radiographics. 2009;29:1811–25.
3. Leyendecker JR, Bloomfeld RS, Di Santis J, et al. MR enterography in the management of patients with Crohn disease. Radiographics. 2009;29:1827–46.
4. Mazziotti S, Ascenti G, Scribano E, et al. Guide to magnetic resonance in Crohn's disease: from common findings to the more rare complicances. Inflamm Bowel Dis. 2011;17:1209–22.
5. Fidler J. MR imaging of the small bowel. In: Carucci LR, editor. Advances in gastrointestinal imaging. Philadelphia: Elsevier Saunders; 2007. p. 317–31.
6. Prassopoulos P, Papanikolaou N, Grammatikakis J, Rousomoustakaki M, Maris T, Gourtsoyiannis N. MR enteroclysis imaging of Crohn disease. Radiographics. 2001;21:Spec No:S161-72. https://doi.org/10.1148/radiographics.21.suppl_1.g01oc02s161. PMID: 11598255
7. Masselli G, Gualdi G. MR imaging of the small bowel. Radiology. 2012;264(2):333–48. https://doi.org/10.1148/radiol.12111658. PMID: 22821694
8. Amzallag-Bellenger E, Oudjit A, Ruiz A, Cadiot G, Soyer PA, Hoeffel CC. Effectiveness of MR enterography for the assessment of small-bowel diseases beyond Crohn disease. Radiographics. 2012;32(5):1423–44. https://doi.org/10.1148/rg.325115088. PMID: 22977028
9. Lohan D, Cronin C, Meehan C, et al. MR small bowel enterography: optimization of imaging timing. Clin Radiol. 2007;62:804–7.
10. Masselli G, Mastroiacovo I, De Marco E, Francione G, Casciani E, Polettini E, Gualdi G. Current tecniques and new perpectives research of magnetic resonance enterography in pediatric Crohn's disease. World J Radiol. 2016;8(7):668–82. https://doi.org/10.4329/wjr.v8.i7.668. PMID: 27551337; PMCID: PMC4965351
11. Mollard BJ, Smith EA, Dillman JR. Pediatric MR enterography: technique and approach to interpretation-how we do it. Radiology. 2015;274(1):29–43. https://doi.org/10.1148/radiol.14122449. PMID: 25531478
12. Siddiki H, Fidler J. MR imaging of the small bowel in Crohn's disease. Eur J Radiol. 2009;69:409–17.
13. Rieber A, Aschoff A, Nussle K, et al. MRI in the diagnosis of small bowel disease: use of positive and negative oral contrast media in combination with enteroclysis. Eur Radiol. 2000;10:1377–82.
14. Laghi A, Paolantonio P, Iafrate F, et al. Oral contrast agents for magnetic resonance imaging of the bowel. Top Magn Reson Imaging. 2002;13:389–96.
15. Karantanas AH, Papanikolau N, Kalef-Ezra J, et al. Blueberry juice used per os in upper abdominal MR imaging: composition and initial clinical data. Eur Radiol. 2000;10:909–13.
16. Kayhan A, Oommen J, Dahi F, et al. Magnetic resonance enterography in Crohn's disease: standard and advanced techniques. World J Radiol. 2010;2:113–21.
17. Biraschi P, Braccini G, Gigoni R, et al. MR enteroclysis using iron oxide particles (ferristene) as an endoluminal contrast agent: an open phase III trial. Magn Reson Imaging. 2004;22:1085–95.
18. Laghi A, Carbone I, Paolantonio P, et al. Polyethylene glycol solution as an oral contrast agent for MR imaging of the small bowel. Acad Radiol. 2002;9(suppl):s355–6.
19. Hermann KA, Zech CJ, Michaely H, et al. Comprehensive magnetic resonance imaging of the small and large bowel using intraluminal dual contrast technique with iron oxide solution and water in magnetic resonance enteroclysis. Investig Radiol. 2005;40:621–9.
20. Schreyer AG, Golder S, Scheibl K, et al. Dark lumen magnetic resonance enteroclysis in combination with MRI colonography for whole bowel assessment in patients with Crohn's disease: first clinical experience. Inflamm Bowel Dis. 2005;11:388–94.
21. Faber SC, Stehling MK, Holzknecht N, et al. Pathologic conditions in the small bowel: findings at fat-suppressed gadolinium-enhanced MR imaging with an optimized suspension of oral magnetic particles. Radiology. 1997;205:278–82.
22. Ajaj W, Goehde SC, Schneemann H, et al. Dose optimization of mannitol solution for small bowel distension in MRI. J Magn Reson Imaging. 2004;20:648–53.
23. Lauenstein TC, Schneemann H, Vogt FM, et al. Optimization of oral contrast agents for MR imaging of the small bowel. Radiology. 2003;228:279–83.
24. Broglia L, Gigante P, Papi C, et al. Magnetic resonance enteroclysis in Crohn's disease. Radiol Med. 2003;106:28–35.
25. Laghi A, Paolantonio P, Catalano C, et al. MR imaging of the small bowel using polyethylene glycol solution as an oral contrast agent in adults and children with celiac disease: preliminary observations. Am J Roentgenol. 2003;180:191–4.
26. Sood RR, Joubert J, Franklin H, et al. Small bowel MRI: comparison of a polyethylene glycol preparation and water as oral contrast media. J Magn Reson Imaging. 2002;15:401–8.
27. McKenna DA, Roche CJ, Murphy JMP, et al. Polyethylene glycol solution as an oral contrast agent for MRI of the small bowel in a patient population. Clin Radiol. 2006;61:966–70.
28. Yacoub HJ, Obara P, Oto A. Evolving role of MRI in Crohn's disease. J Magn Reson Imaging. 2013;37:1277–89.

29. Griffin N, Grant LA, Anderson S, et al. Small bowel MR enterography: problem solving in Crohn's disease. Insights Imaging. 2012;3:251–63.
30. Sinha R, Murphy P, Hawker P, et al. Role of MRI in Crohn's disease. Clin Radiol. 2009;64:341–52.
31. Grand DJ, Beland MD, Machan JT, et al. Detection of Crohn's disease: comparison of CT and MR enterography without antiperistaltic agents performed on the same day. Eur J Radiol. 2012;81:1735–41.
32. Rimola J, Planell N, Rodríguez S, Delgado S, Ordás I, Ramírez-Morros A, Ayuso C, Aceituno M, Ricart E, Jauregui-Amezaga A, Panés J, Cuatrecasas M. Characterization of inflammation and fibrosis in Crohn's disease lesions by magnetic resonance imaging. Am J Gastroenterol. 2015;110(3):432–40. https://doi.org/10.1038/ajg.2014.424. Epub 2015 Jan 27. Erratum in: Am J Gastroenterol. 2015 Mar;110(3):480. PMID: 25623654
33. Bellini D, Rivosecchi F, Panvini N, Rengo M, Caruso D, Carbone I, Ferrari R, Paolantonio P, Laghi A. Layered enhancement at magnetic resonance enterography in inflammatory bowel disease: a meta-analysis. World J Gastroenterol. 2019;25(31):4555–66. https://doi.org/10.3748/wjg.v25.i31.4555. PMID: 31496631; PMCID: PMC6710183
34. Bird ST, Gelperin K, Sahin L, Bleich KB, Fazio-Eynullayeva E, Woods C, Radden E, Greene P, McCloskey C, Johnson T, Shinde M, Krefting I. First-trimester exposure to gadolinium-based contrast agents: a utilization study of 4.6 million U.S. pregnancies. Radiology. 2019;293(1):193–200. https://doi.org/10.1148/radiol.2019190563. Epub 2019 Aug 20. PMID: 31429682
35. Umshaden HW, Szolar D, Grasser J, et al. Small bowel disease: comparison of MR enteroclysis images with conventional enteroclysis and surgical findings. Radiology. 2000;215:717–25.
36. Cicero G, D'Angelo T, Bottari A, Costantino G, Visalli C, Racchiusa S, Marino MA, Cavallaro M, Frosina L, Blandino A, Mazziotti S. Superior mesenteric artery syndrome in patients with Crohn's disease: a description of 2 cases studied with a novel magnetic resonance enterography (MRE) procedure. Am J Case Rep. 2018;19:431–7. https://doi.org/10.12659/ajcr.908273. PMID: 29643328; PMCID: PMC5912011
37. Froelich JM, Waldherr C, Stoupis C, et al. MR motility in Crohn's disease improves lesion detection compared with standard MR imaging. Eur Radiol. 2010;20:1945–51.
38. Odille F, Menys A, Ahmed A, et al. Quantitative assessment of small bowel motility by nonrigid registration of dynamic MR images. Magn Reson Med. 2012;68:783–93.
39. Menys A, Atkinson D, Odille F, et al. Quantified terminal ileal motility during MR enterography as a potential biomarker of Crohn's disease activity: a preliminary study. Eur Radiol. 2012;22:2494–501.
40. Oto A, Kulkarni K, Karczmar GS, et al. Evaluation of diffusion-weighted MR imaging for detection of bowel inflammation in patients with Crohn's disease. Acad Radiol. 2009;16:597–603.
41. Kiryu S, Dodanuki K, Takao H, et al. Free-breathing diffusion weighted imaging for the assessment of inflammatory activity in Crohn's disease. J Magn Reson Imaging. 2009;29:880–6.
42. Cicero G, Mondello S, Wichmann JL, Albrecht MH, Vogl TJ, Cavallaro M, Frosina L, D'Angelo T, Mazziotti S. Fast magnetic resonance Enterography protocol for the evaluation of patients with Crohn's disease: a pilot study. J Clin Imaging Sci. 2020;10:25. https://doi.org/10.25259/JCIS_18_2020. PMID: 32363087; PMCID: PMC7193210
43. Jhaveri KS, Sagheb S, Guimaraes L, Krishna S, Ahari AF, Espin-Garcia O. Evaluation of Crohn disease activity using a potential abbreviated MRE protocol consisting of balanced steady-state free precession MRI only versus full-protocol MRE. AJR Am J Roentgenol. 2021;216(2):384–92. https://doi.org/10.2214/AJR.20.22856. Epub 2020 Dec 9. PMID: 33295814
44. Fiorino G, Bonifacio C, Padrenostro M, Sposta FM, Spinelli A, Malesci A, Balzarini L, Peyrin-Biroulet L, Danese S. Comparison between 1.5 and 3.0 Tesla magnetic resonance enterography for the assessment of disease activity and complications in ileo-colonic Crohn's disease. Dig Dis Sci. 2013;58(11):3246–55. https://doi.org/10.1007/s10620-013-2781-z. Epub 2013 Aug 1. PMID: 23903867
45. Taylor SA, Punwani S, Rodriguez-Justo M, et al. Mural Crohn disease: correlation of dynamic contrast-enhanced MR imaging findings with angiogenesis and inflammation at histologic examination. Pilot Study Radiol. 2009;251:369–79.
46. Del Vescovo R, Sansoni I, Caviglia R, et al. Dynamic contrast enhanced magnetic resonance imaging of the terminal ileum: differentiation of activity of Crohn's disease. Abdom Imaging. 2008;33:417–24.
47. Horsthuis K, Nederveen AJ, de Feiter M, et al. Mapping of T1-values and gadolinium-concentrations in MRI as indicator of disease activity in luminal Crohn's disease: a feasibility study. J Magn Reson Imaging. 2009;29:488–93.
48. Giusti S, Faggioni L, Neri E, et al. Dynamic MRI of the small bowel: usefulness of quantitative contrast-enhancement parameters and time-signal intensity curves for differentiating between active and inactive Crohn's disease. Abdom Imaging. 2010;35:646–53.
49. Oto A, Fan X, Mustafi D, et al. Quantitative analysis of dynamic contrast enhanced MRI for assessment of bowel inflammation in Crohn's disease: pilot study. Acad Radiol. 2009;16:1223–30.
50. Ippolito D, Lombardi S, Talei Franzesi C, Drago SG, Querques G, Casiraghi A, Pecorelli A, Riva L, Sironi S. Dynamic contrast-enhanced MR with quantitative perfusion analysis of small bowel in vascular assessment between inflammatory and fibrotic lesions in Crohn's disease: a feasibility study. Contrast Media Mol Imaging. 2019;2019:1767620. https://

doi.org/10.1155/2019/1767620. PMID: 30863218; PMCID: PMC6378774
51. Vieujean S, Coibion C, Seidel L, Louis E, Meunier P. Magnetic resonance enterography perfusion parameters reveal complex changes in affected and unaffected segments in Crohn's disease. Scand J Gastroenterol. 2020;55(9):1041–8. https://doi.org/10.1080/00365521.2020.1802773. Epub 2020 Aug 6. PMID: 32757858
52. Bufman H, Eliakim R, Tau N, Amitai MM. Magnetic resonance enterography in Crohn's disease patients: current state of the art and future perspectives. Expert Rev Med Devices. 2021;18(7):657–67. https://doi.org/10.1080/17434440.2021.1939682. Epub 2021 Jul 30. PMID: 34109891
53. Park SH, Ye BD, Lee TY, Fletcher JG. Computed tomography and magnetic resonance small bowel enterography: current status and future trends focusing on Crohn's disease. Gastroenterol Clin N Am. 2018;47(3):475–99. https://doi.org/10.1016/j.gtc.2018.04.002. Epub 2018 Jul 7. PMID: 30115433
54. Li Y, Hauenstein K. New imaging techniques in the diagnosis of inflammatory bowel diseases. Viszeralmedizin. 2015;31(4):227–34. https://doi.org/10.1159/000435864. Epub 2015 Jul 28. PMID: 26557830; PMCID: PMC4608604
55. Adler J, Swanson SD, Schmiedlin-Ren P, Higgins PD, Golembeski CP, Polydorides AD, McKenna BJ, Hussain HK, Verrot TM, Zimmermann EM. Magnetization transfer helps detect intestinal fibrosis in an animal model of Crohn disease. Radiology. 2011;259(1):127–35. https://doi.org/10.1148/radiol.10091648. Epub 2011 Jan 28. PMID: 21324841; PMCID: PMC3064818
56. Dillman JR, Swanson SD, Johnson LA, Moons DS, Adler J, Stidham RW, Higgins PD. Comparison of noncontrast MRI magnetization transfer and T2-weighted signal intensity ratios for detection of bowel wall fibrosis in a Crohn's disease animal model. J Magn Reson Imaging. 2015;42(3):801–10. https://doi.org/10.1002/jmri.24815. Epub 2014 Dec 13. PMID: 25504823
57. Li XH, Mao R, Huang SY, Sun CH, Cao QH, Fang ZN, Zhang ZW, Huang L, Lin JJ, Chen YJ, Rimola J, Rieder F, Chen MH, Feng ST, Li ZP. Characterization of degree of intestinal fibrosis in patients with Crohn disease by using magnetization transfer MR imaging. Radiology. 2018;287(2):494–503. https://doi.org/10.1148/radiol.2017171221. Epub 2018 Jan 19. PMID: 29357272; PMCID: PMC6911697
58. Pazahr S, Blume I, Frei P, Chuck N, Nanz D, Rogler G, Patak M, Boss A. Magnetization transfer for the assessment of bowel fibrosis in patients with Crohn's disease: initial experience. MAGMA. 2013;26(3):291–301. https://doi.org/10.1007/s10334-012-0355-2. Epub 2012 Nov 9. PMID: 23138635
59. Yoon K, Chang KT, Lee HJ. MRI for Crohn's disease: present and future. Biomed Res Int. 2015;2015:786802. https://doi.org/10.1155/2015/786802. Epub 2015 Aug 27. PMID: 26413543; PMCID: PMC4564596
60. Kohli A, Holzwanger EA, Levy AN. Emerging use of artificial intelligence in inflammatory bowel disease. World J Gastroenterol. 2020;26(44):6923–8. https://doi.org/10.3748/wjg.v26.i44.6923. PMID: 33311940; PMCID: PMC7701951
61. Arkko A, Kaseva T, Salli E, Mäkelä T, Savolainen S, Kangasniemi M. Automatic detection of Crohn's disease using quantified motility in magnetic resonance enterography: initial experiences. Clin Radiol. 2022;77(2):96–103. https://doi.org/10.1016/j.crad.2021.10.006. Epub 2021 Nov 6. PMID: 34753588
62. Naziroglu RE, Puylaert CAJ, Tielbeek JAW, Makanyanga J, Menys A, Ponsioen CY, Hatzakis H, Taylor SA, Stoker J, van Vliet LJ, Vos FM. Semi-automatic bowel wall thickness measurements on MR enterography in patients with Crohn's disease. Br J Radiol. 2017;90(1074):20160654. https://doi.org/10.1259/bjr.20160654. Epub 2017 May 23. PMID: 28401775; PMCID: PMC5602169
63. Tabari A, Kilcoyne A, Jeck WR, Mino-Kenudson M, Gee MS. Texture analysis of magnetic resonance Enterography contrast enhancement can detect fibrosis in Crohn disease strictures. J Pediatr Gastroenterol Nutr. 2019;69(5):533–8. https://doi.org/10.1097/MPG.0000000000002454. PMID: 31365485
64. Makanyanga J, Ganeshan B, Rodriguez-Justo M, Bhatnagar G, Groves A, Halligan S, Miles K, Taylor SA. MRI texture analysis (MRTA) of T2-weighted images in Crohn's disease may provide information on histological and MRI disease activity in patients undergoing ileal resection. Eur Radiol. 2017;27(2):589–97. https://doi.org/10.1007/s00330-016-4324-4. Epub 2016 Apr 5. PMID: 27048528; PMCID: PMC5209452

MR Enterography: Anatomy

4

Tommaso D'Angelo, Giorgio Ascenti, Silvio Mazziotti, and Giuseppe Cicero

4.1 Normal MR Anatomy of Duodenum and Small Bowel

The small bowel consists of duodenum, jejunum, and ileum. It begins at the pylorus of the stomach and ends at the ileocecal junction (Fig. 4.1) [1, 2].

It is called small bowel because the luminal caliber is smaller than that of the colon, although it is longer in length (6 meters on average) [3].

The small intestine is differentiated from the large bowel also due to the presence of a mesentery (exceptions being no mesentery in the duodenum and mesentery in the transverse and sigmoid colon) and the absence of taenia coli and appendices epiploicae.

The duodenum begins at the pylorus on the right-hand side of the upper abdominal cavity and ends at the duodenojejunal junction at the level of ligament of Treitz. Anatomically, four segments can be recognized within the duodenum: superior, descending, horizontal, and ascending. Together, these parts form a C-shaped structure.

The jejunum and ileum are the two distal parts of small intestine. Since there is not a clear separation between the two segments, they are often indicated as the jejunoileum.

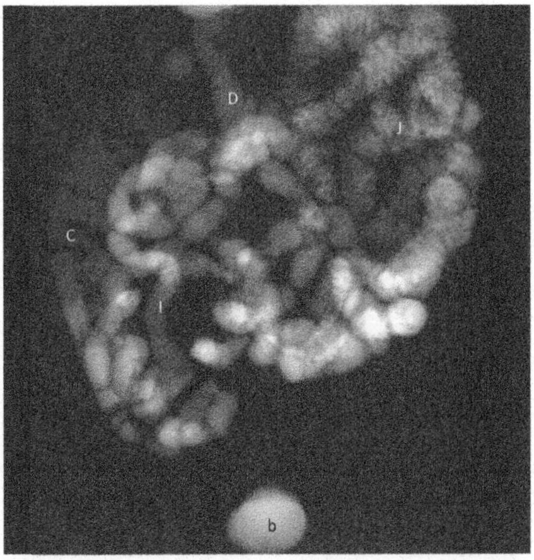

Fig. 4.1 Coronal single-shot thick-slab T2-w RARE. A panoramic view of the small bowel. **D**: duodenum; **J**: jejunum; **I**: ileum; **B**: bladder; **C**: colon

On MR images, the duodenal "C-shaped" as well as the duodenal and jejunal wall, lumen and fold patterns are well demonstrated. The small bowel diameter normally should not exceed 3 cm and shows a progressive decrease from the duodenum (approximately 25 mm) to the terminal ileum (19 mm) (Fig. 4.2) [3, 4].

Also the number of folds is reduced passing from the duodenum (approximately 4.5 per 2.5 cm) to terminal ileum (1.5 per 2.5 cm) (Figs. 4.3, 4.4 and 4.5) [4].

T. D'Angelo · G. Ascenti · S. Mazziotti
G. Cicero (✉)
Department of Biomedical Sciences and Morphological and Functional Imaging, University Hospital Messina, Messina, Italy
e-mail: tdangelo@unime.it; gascenti@unime.it; smazziotti@unime.it; gcicero@unime.it

S. Mazziotti et al. (eds.), *MR Enterography*, https://doi.org/10.1007/978-3-031-11930-9_4

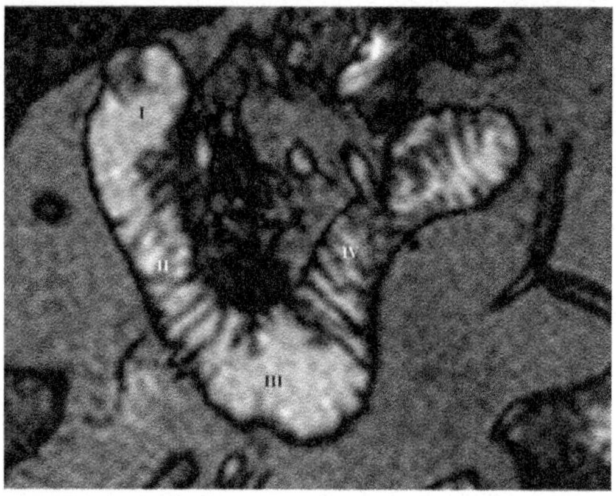

Fig. 4.2 Particular of coronal True-FISP image. Different parts of duodenum, which appears distended and with its characteristic fold pattern. **I**: superior part; **II**: descending part; **III**: horizontal part; **IV**: ascending part

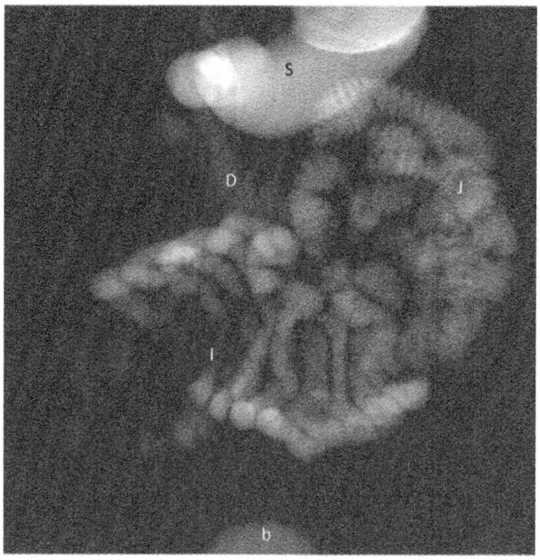

Fig. 4.3 Coronal single-shot thick-slab T2-w RARE. Note how duodenum (**D**) and jejunum (**J**) have comparable bowel diameter and fold thickness, similar number of folds and interfold distances. **S**: stomach; **I**: ileum; **B**: bladder

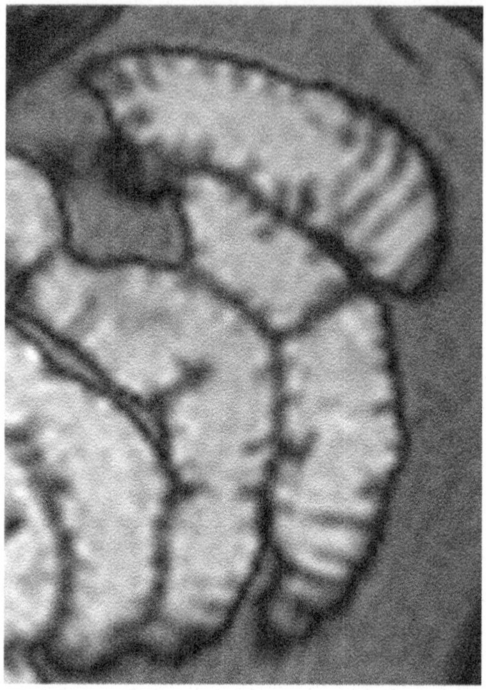

Fig. 4.4 Magnified coronal True-FISP image of jejunum where it is well demonstrated the normal mucosal, mural, and fold pattern

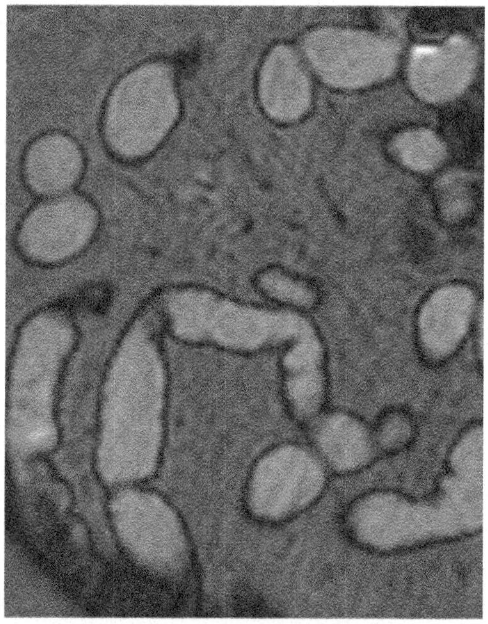

Fig. 4.5 Magnified coronal True-FISP image of ileum with normal mucosal and mural aspect. Note how fewer folds are visualized in comparison to those of jejunum and duodenum, as seen in the previous figures

Normal bowel wall thickness, when adequately distended, should not overcome the threshold of 3 mm [5].

As previously stated, due to significant improvements in gradient technology, the visualization of the mesenteric structures has become possible through the use of True-FISP sequence. Very small mesenteric lymph nodes and vessels can be consistently detected avoiding motion artifacts due to the short acquisition time [6, 7].

4.2 Normal MR Anatomy of Sphincters and Perianal Region

The anal canal is the terminal portion of the digestive tract and it is circumscribed by the rectum above and the anal verge within the perineum below. It is essentially a cylindrical structure with length variable between 3 and 5 cm, with two-thirds of this lying above and one-third below the dentate line. The anal canal is completely extraperitoneal virtual cavity delimited by the internal and external sphincters [8–10].

The internal sphincter is the distal termination of the circular muscle of the gut tube and contributes to forming the inner surface of the canal. It is composed of concentric layers of smooth circular involuntary muscular fibers and it is responsible for 85% of resting anal tone [10].

On the other hand, the external sphincter is a layer of voluntary striated muscle encircling the anal canal and anal verge.

The external sphincter has its posterior attachment to the anococcygeal ligament and anterior attachment to the perineal body and urogenital diaphragm; it merges proximally with the sling-like puborectalis muscle which in turn merges with the levator plate of the pelvic floor. The external sphincter contributes only 15% of resting anal tone, but its strong voluntary contractions can prevent defecation [10]. Cutting of the external sphincter can determine incontinence.

There is an adipose interstice between the internal and external anal sphincters, known as the intersphincteric space.

The ischioanal fossae lie laterally to the sphincteric complex and contain a mixture of fat and fibroelastic connective tissue. Cranially, the ischioanal fossae are separated from the ischiorectal ones by the levator ani muscle (Fig. 4.6) [9].

All these structures, due to the high contrast resolution, are well depicted on both coronal and axial MR images (Fig. 4.7) [11].

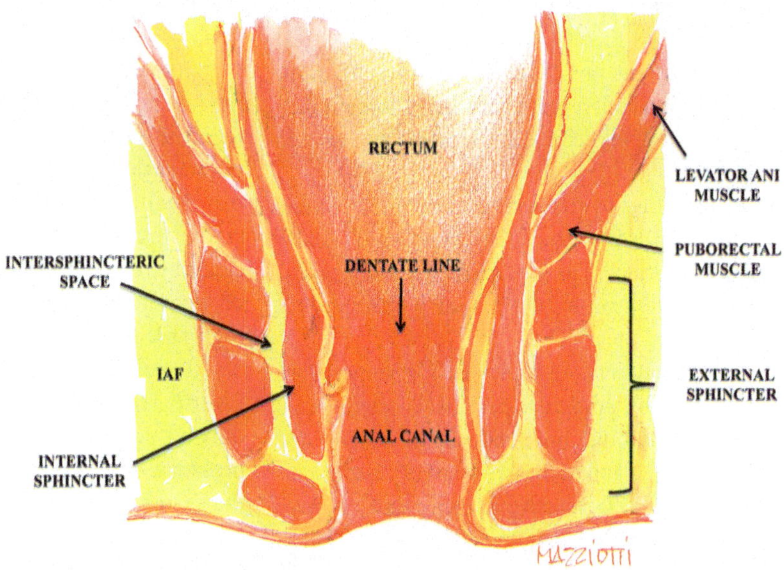

Fig. 4.6 Illustration showing the normal anatomy of the anal canal and perianal region along the coronal plane. **IAF**: ischioanal fossa

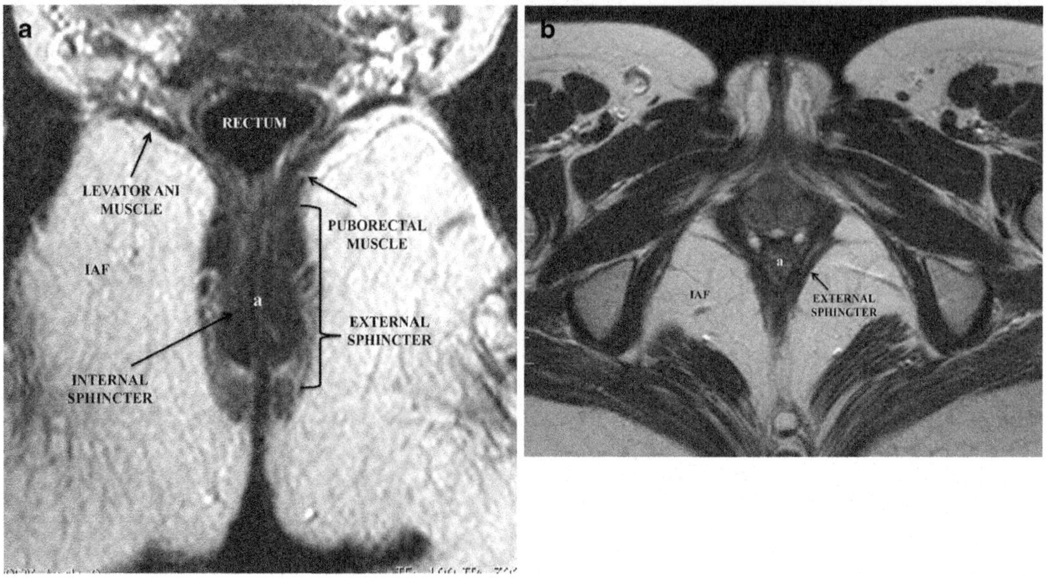

Fig. 4.7 Coronal-oblique (**a**) and axial-oblique (**b**) T2-w TSE images showing the normal MR anatomy of perianal region. **a**: anal canal; **IAF**: ischioanal fossa

4.3 Anatomical Changes after Surgery: Types of Interventions and Anastomoses

4.3.1 Small and Large Bowel

In IBD patients suffering from active inflammation, surgery has to be taken into consideration after failure of medical treatment [12, 13].

Interventions in CD patients can ensure clinical remission over a period of time, although they do not definitively cure CD as generally occur to UC patients.

In fact, due to the recurrent behavior of the disease, multiple intestinal operations are usually required during the patient's lifetime, thus exposing to the risk of a short bowel syndrome [12, 13].

In CD patients, risk factors for surgery are represented by smoking, perianal disease, presence of granuloma, high endoscopic scores, and high values of inflammatory markers (stool calprotectin level, C-reactive protein, and erythrocyte sedimentation) [14].

The indications for surgery are represented by acute disease complications, such as toxic colitis with or without associated megacolon, hemorrhage, and perforation, as well as chronic complications, like extraintestinal manifestations (i.e., fistulas, abscesses) or malignant degeneration [13, 15].

Surgery encompasses resectional and nonresectional approaches, the latter including internal bypass and strictureplasty.

Internal bypass historically represented one of the first procedures employed in CD patients and requires oversewn of the pathologic segment with anastomosis fashioning between the proximal health intestinal loop and the transverse colon. In a second time, the pathologic segment could be restored, if healed after medical therapy, or definitively resected [16]. Considering the high risk of malignant degeneration of the diseased intestinal segment, this approach fell out of favor over the years and it is currently employed only in specific clinical scenarios, such as the presence of ileocecal phlegmon involving the iliac vessels and/or retroperitoneum or upper gastrointestinal tract localization, needing an extensive surgical reconstruction of the pancreaticobiliary system [16].

Strictureplasty is a surgical approach whose main purpose is to avoid intestinal resection [12, 16–18].

It requires an incision along the antimesenteric border of the narrowed loop, creating an enterotomy which is then sewn transversely (i.e., Heineke-Mikulicz technique) or through a side-to-side anastomosis (i.e., Finney technique).

Different types of strictureplasty exist and are chosen on the basis of the length (Heineke-Mikulicz if <10 cm; Finney if 10–25 cm; Michelassi >20 cm) and according to the pathologic features of the affected loop [16, 18].

Strictureplasty is suggested in case of diffuse involvement of the small bowel with multiple strictures, previous intestinal resection, and short bowel syndrome, whereas it is contraindicated when the involved loop is affected by phlegmonous inflammation, internal and/or external fistulas or in case of perforation [12, 16].

Nonetheless, strictureplasty is not recommended for duodenal or colonic CD lesions, due to lower effectiveness, or in case of an ileocecal valve involvement, in which resection should be preferred [12, 19].

Intestinal resection requires removal of the pathologic loop with fashioning of a communication between the upstream and the downstream healthy segments through an anastomosis.

Different types of surgical techniques are available and can be employed on CD patients (Fig. 4.8).

The laparoscopic approach takes advantage from the lower hospitalization and postoperative complications and better esthetic results [12].

On the other hand, laparotomy is usually performed in case of emergency or presence of fistulas or phlegmons.

Up to now, side-to-side stapled anastomosis are generally preferred to hand-sewn end-to-end anastomosis due to lower rates of postoperative complications and surgical recurrences [19, 20].

The "Kono-S" is a novel type of side-to-side hand-sewn anastomosis which requires two stapled transection ends with isolation of the anastomosis from the mesentery in order to decrease risk of inflammation and relapse [13, 20].

So far, Kono-S anastomosis has shown encouraging results in terms of lower surgical recurrences, postoperative complications, and higher surgery-free survival rates [13, 16, 20–22].

When colonic involvement is extensive and the rectum is spared, the ileum is sutured to the anal canal configuring an ileal pouch-anal anastomosis (IPAA) which can assume variable appearances (i.e., "J," "S," "W," or "H") (Fig. 4.9).

Ileal-anal pouches are preferably performed in UC patients than in CD ones, due to the higher risk of inflammations within the latter [18, 21].

On the other hand, when proctectomy is necessary, due to medical treatment failure, presence of perianal fistulas and/or abscesses, and fecal incontinence, fecal diversion (or ostomy) is performed [12].

It requires the connection of an intestinal segment to an external bag in order to recover the distal pathologic intestinal segment (Fig. 4.10) [18].

Cutaneous ostomy may be complicated by stenosis, parastomal hernias, and soaking of parastomal surface [18].

Nevertheless, this type of intervention can be intended also as a temporary approach for the time of disease healing of the former pathologic segment.

Healing rates seem to be accelerated when ostomy is associated to immunosuppressive or biological therapy [12].

4.3.2 Anal Canal

The anorectal district is involved in 30% of patients affected by CD and it is estimated that 20% of these patients will need surgical operation during their lifetime [12, 19].

Pathologic conditions in this area include fissures, skin tags, abscesses or fistulas, strictures, or malignancies.

Surgical intervention may represent an alternative or can be performed in combination with medications.

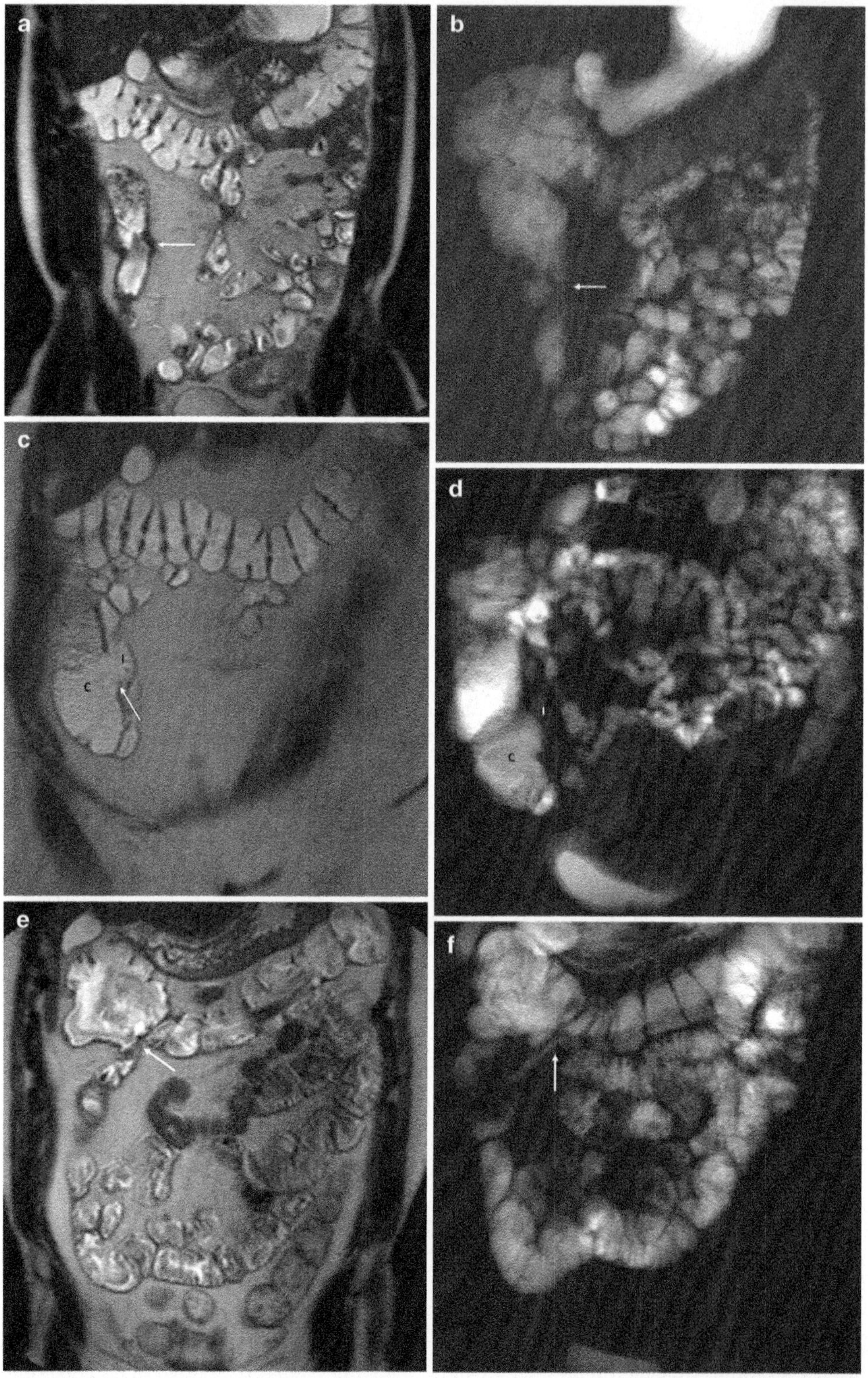

Fig. 4.8 Examples of surgical resections and ileocolic anastomoses in CD patients. Coronal T2-w TSE (**a**) and coronal RARE T2-wthick-slab (**b**) acquisitions showing a terminoterminal ileocolic anastomosis (arrow). Coronal T2-w TSE (**c**) and coronal RARE T2-w thick-slab (**d**) images demonstrate a lateroterminal ileocolic anastomosis (arrow); **I**: ileum; **C**: colon. Coronal T2-w TSE (**e**) and coronal T2-w RARE thick-slab (**f**) scans display an ileo-transverse anastomosis (arrow)

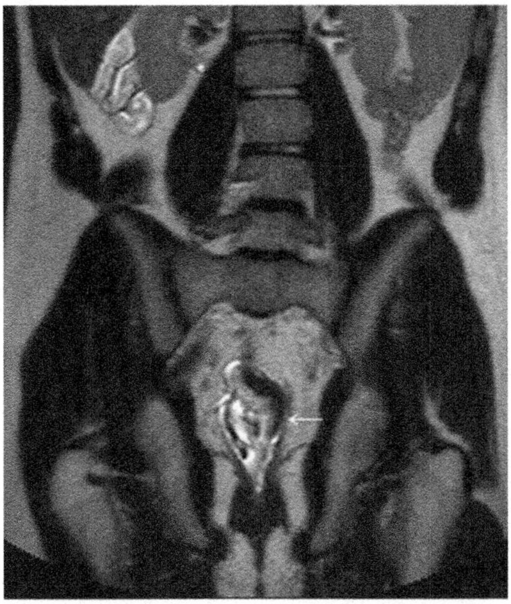

Fig. 4.9 Coronal T2-w HASTE shows ileoanal pouch (arrow) constructed after colectomy

Fistulotomy is suitable for treatment of simple and superficial fistulas and requires incision and curette of the track with secondary intention wound healing [23, 24].

Contraindications to fistulotomy include complex fistulas, wide sphincter involvement, presence of active proctitis and abscesses [23, 25].

Fistulotomy is not recommended in case of anal sphincter involvement or for anterior fistulas, due to the high risk of fecal incontinence [12]. In these cases, setons and medical therapy are recommended.

Non-cutting setons are positioned within the track in order to drain the fistula and avoid the risk of abscess formation.

Setons are generally used in case of active proctitis or for avoiding surgery [26, 27].

There is no established optimal timing for their removal, ranging from few weeks to several months or years [27].

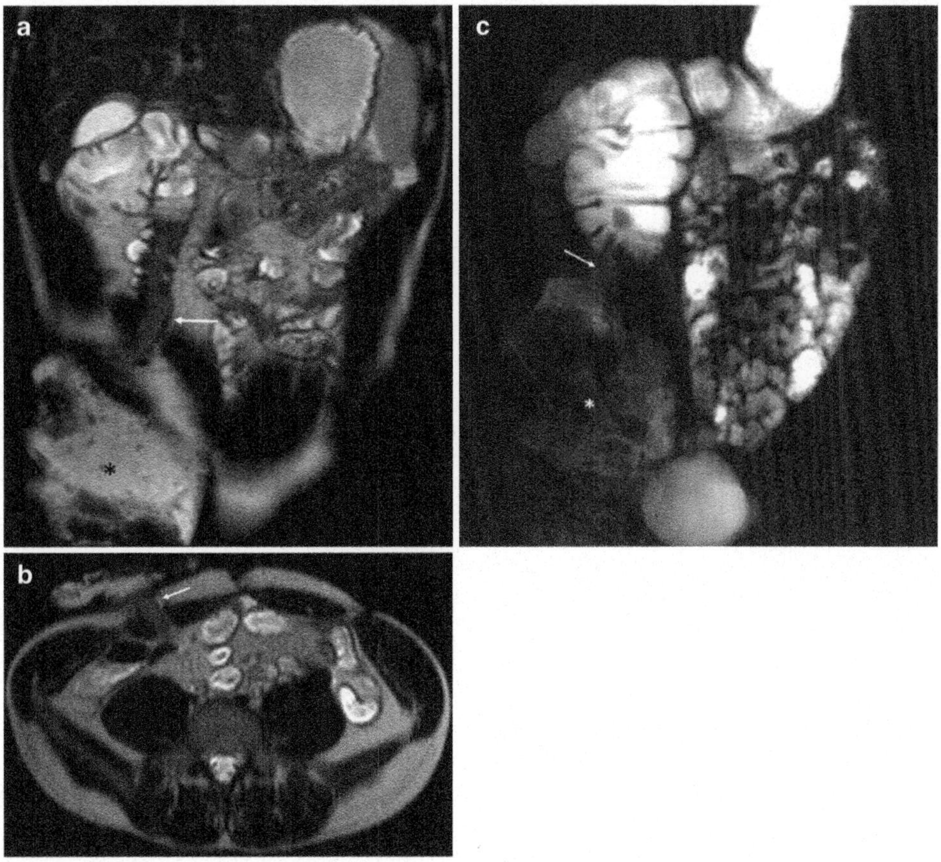

Fig. 4.10 Coronal (**a**) and axial (**b**) T2-w HASTE acquisitions together with coronal RARE T2-w thick-slab scan (**c**) well demonstrate a cutaneous stoma (arrow) within the right flank and the external colostomy bag (asterisk)

Effectiveness of setons can be increased by association with immunosuppressive and biologic therapies [23, 27].

Among the various surgical options, some are less invasive and can preserve the sphincteric structures.

One is the "endorectal advancement flap" [12].

This technique requires the placement of a U-shaped endorectal flap, consisting in mucosal, submucosal, or muscular layer, in order to cover the internal opening of the fistula [12, 24, 27, 28].

Success rates are variable and especially low in CD patients with extensive small bowel involvement, proctitis, and anal stenosis [23, 27].

Although in a small percentage of cases (9–10%), this procedure exposes the patient to the risk of fecal incontinence [28].

Other "sphincter-sparing" approaches are represented by use of in-fill materials, such as fibrin glue and fistula plug.

The former requires an "intra-fistula" injection of a mixture of fibrinogen and thrombin [12, 23, 25].

Instead, anal fistula plugs are bioprosthetic devices whose main purpose is to obstruct the fistula tracks and promote healing [25, 28].

The performance of fibrin glue or fistula plug remains controversial, mainly due to the small number of studies and the heterogeneous results in terms of healing rates [12, 23, 26–28].

Ligation of Intersphincteric Fistula Tract (LIFT) procedure requires the interruption and removal of the fistulous track at the level of intersphincteric groove, and ligation of the internal and external sphincter sides [23, 24, 26].

Although so far this sphincter-sparing technique has been mainly performed on non-CD subjects, it is suitable for CD patients without proctitis and coexistent active small bowel inflammation [23, 25, 27].

Video Assisted Anal Fistula Treatment (VAAFT) is a procedure consisting in real-time appreciation of the fistulous path through a fistuloscope and subsequent cauterization with closure of the internal opening by simple sutures or staples.

On the other hand, destruction and closure of the fistula can be also performed using laser energy [25].

Although the undoubted advantages of short hospitalization and sphincter maintenance, the main drawback of this procedure remains the lack of a direct visualization of the fistulous tack [26, 27].

Outcomes could be improved by previous positioning of a seton or subsequent internal opening closure through advancement flap [26, 27].

A novel and fascinating approach for fistula healing and repair is represented by injection of stem cells that are provided with anti-inflammatory and immunomodulatory activity [26].

Resort to invasive surgery becomes necessary after failure of other treatment options.

Fecal diversion represents a final attempt for promoting healing of the anal and perianal structures and requires a cutaneous stoma for deviation of stool passage [23].

In case of success, intestinal continuity will be restored, otherwise proctectomy becomes mandatory [12, 23, 24].

Complications of perianal disease in CD patients include abscess formation, strictures, and malignancies.

Antibiotics alone may be the only therapy in case of small abscesses (1 cm) [25].

Perianal abscesses can be treated through the positioning of a drainage catheter in combination with antibiotic therapy [25].

On the other hand, pelvic abscesses can be treated by surgery or drainage through endoscopy and intraluminal pig-tail stent positioning [25].

Strictures are caused by fibrosis as a consequence of alternation of inflammatory and reparative processes.

Minimally aggressive methods include balloon dilation or use of intra-anal dilators [23].

In refractory cases, stricturotomy or proctectomy is performed [23, 25].

Risk of anal malignancies is higher in CD patients.

Anal tumors are usually represented by squamous cell carcinoma or adenocarcinoma.

Tumor resection is the same performed in non-CD patients and may require perioperative adjuvant therapy as well as lymphadenectomy [19].

References

1. Mazziotti S, Ascenti G, Scribano E, Gaeta M, Pandolfo A, Bombaci F, Donato R, Fries W, Blandino A. Guide to magnetic resonance in Crohn's disease: from common findings to the more rare complicances. Inflamm Bowel Dis. 2011;17(5):1209–22. https://doi.org/10.1002/ibd.21548. Epub 2010 Nov 5
2. Cicero G, Mazziotti S. Crohn's disease at radiological imaging: focus on techniques and intestinal tract. Intest Res. 2021;19(4):365–78. https://doi.org/10.5217/ir.2020.00097. Epub 2020 Nov 25. PMID: 33232590; PMCID: PMC8566824
3. Laghi A, Hara AK. Small bowel disease. In: Hodler J, Kubik-Huch RA, von Schulthess GK, editors. Diseases of the abdomen and pelvis 2018–2021: diagnostic imaging—IDKD book; 2018. Mar 21.
4. Cronin C, Delappe E, Lohan DG, et al. Normal small bowel wall characteristics on MR enterography. Eur J Radiol. 2010;75:207–11.
5. Griffin N, Grant LA, Anderson S, Irving P, Sanderson J. Small bowel MR enterography: problem solving in Crohn's disease. Insights Imaging. 2012;3(3):251–63. https://doi.org/10.1007/s13244-012-0154-3. Epub 2012 Mar 17. PMID: 22696087; PMCID: PMC3369125
6. Kayhan A, Oommen J, Dahi F, et al. Magnetic resonance enterography in Crohn's disease: standard and advanced techniques. World J Radiol. 2010;2:113–21.
7. Furukawa A, Saotome T, Yamasaki M, et al. Cross sectional imaging in Crohn disease. Radiographics. 2004;24:689–702.
8. Horsthius K, Stoker J. MRI of perianal Crohn's disease. Am J Roentgenol. 2004;183:1309–15.
9. Halligan S, Stoker J. Imaging of fistula in ano. Radiology. 2006;239:18–33.
10. Morris J, Spencer JA, Ambrose S, et al. MR imaging classification of perianal fistulas and its implications for patient management. Radiographics. 2000;20:623–35.
11. Cicero G, Ascenti G, Blandino A, Pallio S, Abate C, D'Angelo T, Mazziotti S. Magnetic resonance imaging of the anal region: clinical applications. J Clin Imaging Sci. 2020;10:76. https://doi.org/10.25259/JCIS_180_2020. PMID: 33274120; PMCID: PMC7708963
12. Seifarth C, Kreis ME, Gröne J. Indications and specific surgical techniques in Crohn's disease. Viszeralmedizin. 2015;31(4):273–9. https://doi.org/10.1159/000438955. Epub 2015 Aug 14. PMID: 26557836; PMCID: PMC4608647
13. Ingallinella S, Campanelli M, Antonelli A, Arcudi C, Bellato V, Divizia A, Franceschilli M, Petagna L, Sensi B, Sibio S, Siragusa L, Sica GS. The role of active inflammation and surgical therapy in Crohn's disease recurrence. Gastroenterol Res Pract. 2020;2020:2845407. https://doi.org/10.1155/2020/2845407. PMID: 33456458; PMCID: PMC7785378
14. El Megeed KHA, Saleh SAB, Mohamed AE, Anwar CA. Predictors of surgical intervention in patients with inflammatory bowel disease. Egypt J Intern Med. 2021;33:19.
15. Michelassi F, Sultan S. Surgical treatment of complex small bowel Crohn disease. Ann Surg. 2014;260(2):230–5. https://doi.org/10.1097/SLA.0000000000000697.
16. Meima-van Praag EM, Buskens CJ, Hompes R, Bemelman WA. Surgical management of Crohn's disease: a state of the art review. Int J Color Dis. 2021;36(6):1133–45. https://doi.org/10.1007/s00384-021-03857-2. Epub 2021 Feb 2. PMID: 33528750; PMCID: PMC8119249
17. Bemelman WA, Warusavitarne J, Sampietro GM, Serclova Z, Zmora O, Luglio G, de Buck van Overstraeten A, Burke JP, Buskens CJ, Colombo F, Dias JA, Eliakim R, Elosua T, Gecim IE, Kolacek S, Kierkus J, Kolho KL, Lefevre JH, Millan M, Panis Y, Pinkney T, Russell RK, Shwaartz C, Vaizey C, Yassin N, D'Hoore A. ECCO-ESCP consensus on surgery for Crohn's disease. J Crohns Colitis. 2018;12(1):1–16. https://doi.org/10.1093/ecco-jcc/jjx061.
18. Cicero G, Ascenti G, Blandino A, Trimarchi R, Booz C, Vogl TJ, D'Angelo T, Mazziotti S. Elective surgery outcomes in inflammatory bowel disease: interpretation at magnetic resonance enterography. Jpn J Radiol. 2021;39(7):633–41. https://doi.org/10.1007/s11604-021-01103-x. Epub 2021 Feb 24
19. Toh JW, Stewart P, Rickard MJ, Leong R, Wang N, Young CJ. Indications and surgical options for small bowel, large bowel and perianal Crohn's disease. World J Gastroenterol. 2016;22(40):8892–904. https://doi.org/10.3748/wjg.v22.i40.8892. PMID: 27833380; PMCID: PMC5083794
20. Reynolds IS, Doogan KL, Ryan ÉJ, Hechtl D, Lecot FP, Arya S, Martin ST. Surgical strategies to reduce postoperative recurrence of Crohn's disease after ileocolic resection. Front Surg. 2021;17(8):804137. https://doi.org/10.3389/fsurg.2021.804137. PMID: 34977147; PMCID: PMC8718441
21. Kono T, Ashida T, Ebisawa Y, et al. A new antimesenteric functional end-to-end handsewn anastomosis: surgical prevention of anastomotic recurrence in Crohn's disease. Dis Colon Rectum. 2011;54(5):586–92.
22. Luglio G, Rispo A, Imperatore N, et al. Surgical prevention of anastomotic recurrence by excluding mesentery in Crohn's disease: the SuPREMe-CD study—a randomized clinical trial. Ann Surg. 2020;272(2):210–7.
23. Williams JL, Shaffer VO. Modern management of perianal Crohn's disease: a review. Am Surg. 2021;87(9):1361–7. https://doi.org/10.1177/0003134820956331. Epub 2020 Dec 19

24. Kelley KA, Kaur T, Tsikitis VL. Perianal Crohn's disease: challenges and solutions. Clin Exp Gastroenterol. 2017;10:39–46. https://doi.org/10.2147/CEG.S108513. PMID: 28223835; PMCID: PMC5308478
25. Kotze PG, Shen B, Lightner A, Yamamoto T, Spinelli A, Ghosh S, Panaccione R. Modern management of perianal fistulas in Crohn's disease: future directions. Gut. 2018;67(6):1181–94. https://doi.org/10.1136/gutjnl-2017-314918. Epub 2018 Jan 13
26. Zabot GP, Cassol O, Saad-Hossne R, Bemelman W. Modern surgical strategies for perianal Crohn's disease. World J Gastroenterol. 2020;26(42):6572–81. https://doi.org/10.3748/wjg.v26.i42.6572. PMID: 33268947; PMCID: PMC7673971
27. Gold SL, Cohen-Mekelburg S, Schneider Y, Steinlauf A. Perianal fistulas in patients with Crohn's disease, part 2: surgical, endoscopic, and future therapies. Gastroenterol Hepatol (N Y). 2018;14(9):521–8. PMID: 30364296; PMCID: PMC6194657
28. Adegbola SO, Pisani A, Sahnan K, Tozer P, Ellul P, Warusavitarne J. Medical and surgical management of perianal Crohn's disease. Ann Gastroenterol. 2018;31(2):129–39. https://doi.org/10.20524/aog.2018.0236. Epub 2018 Feb 8. PMID: 29507460; PMCID: PMC5825943

5 MR-Enterography Intestinal Findings in Crohn's Disease

Giuseppe Cicero, Alfredo Blandino, Tommaso D'Angelo, and Silvio Mazziotti

Inflammatory bowel disease (IBD) primarily refers to two chronic diseases that cause chronic inflammation of all or part of the digestive tract: ulcerative colitis (UC) and Crohn's disease (CD).

Despite a sporadic involvement of the terminal ileum as a primary localization or due to backwash ileitis, UC predominantly affects the large bowel. Colonoscopy is therefore the most effective method to assess disease activity.

On the other hand, CD is an idiopathic, chronic, transmural, inflammatory disease that can affect any part of the gastrointestinal tract from mouth to anus, with a tendency toward segmental distribution and often involving multiple discontinuous sites (the so-called "skip lesions").

In about 70–80% of patients with CD the small bowel is affected and in about 20–30% the disease is limited to the small bowel. The colon can be affected either with (50% of cases) or without (15–20%) the involvement of the small intestine [1].

The incidence of CD seems to have a bimodal distribution, the first peak occurring in late adolescence and early adulthood while a second smaller increase in incidence can be seen between the fifth and seventh decade of life [2, 3].

The etiology and pathogenesis of CD are not completely understood, but there is increasing evidence that genetic as well as environmental factors may play an important role in causing a sustained activation of mucosal immune responses, which lead to cytokine overproduction and subsequently to leukocytic infiltration of the bowel wall [2, 4]. Furthermore, intestinal microbiota would also likely play a major role in the pathogenesis of CD.

Microscopically, the initial lesion starts as a focal inflammatory infiltrate in the mucosa and submucosa, which then leads to hyperemia and edema. Macroscopically, the earliest feature of CD is a shallow aphthoid mucosal ulcer, histologically corresponding to initial mucosal ulceration over a mucosal lymphoid follicle. As the disease progresses, aphthoid ulcers develop and further progress into extensive linear ulcers and fissures. Advanced ulcerations with bulging of the edematous residual mucosal islands produce the typical ulceronodular or "cobblestone" appearance. In more severe cases, transmural inflammation and serosal involvement are present as well.

The bowel wall can also result to be thickened due to the coexistence of inflammatory infiltrates and fibrosis. In long-standing cases, chronic

G. Cicero · A. Blandino · T. D'Angelo
S. Mazziotti (✉)
Department of Biomedical Sciences and Morphological and Functional Imaging, University Hospital Messina, Messina, Italy
e-mail: gcicero@unime.it; ablandino@unime.it; tdangelo@unime.it; smazziotti@unime.it

S. Mazziotti et al. (eds.), *MR Enterography*, https://doi.org/10.1007/978-3-031-11930-9_5

obstruction due to scarring, luminal narrowing, and strictures may arise. For what concerns extramural manifestations, these are fistulas, abscesses, adhesions, creeping fat, and enlargement of the lymph nodes.

Although not frequent, toxic megacolon and neoplasms such as lymphoma and carcinoma may also occur [5].

Symptoms of CD are often unpleasant and they are classically represented by abdominal pain, weight loss, diarrhea (which may include blood or mucus), and tenesmus; however, it can have different presentations and tends to have an unpredictable course marked by flares, remissions, and relapses.

The diagnosis of CD is based on a combination of clinical findings, endoscopic appearance, biopsy, radiological studies, and biochemical markers; however, most commonly, ileocolonoscopy and biopsies from the terminal ileum and colon are the ones used to achieve the final diagnosis.

With regard to the imaging of CD at initial presentation and in the setting of clinical suspicion, as previously stated in the introduction, MR of the small bowel can be considered a useful support both in patients with an incomplete ileocolonoscopy and in those with negative ileocolonoscopy, to, respectively, exclude enteric inflammation at the level of the terminal ileum or proximal to this site.

MR of the small bowel can also be used as a first-line diagnostic approach in pediatric patients with clinical suspicion for CD, or it can be performed prior to video capsule endoscopy (very sensitive to detect subtle mucosal disease), to exclude strictures.

However, in the majority of cases, MRE is primarily employed to assess the exact location and extent of inflammatory bowel disease in patients with histological diagnosis of CD.

It should also be considered that in clinical practice, CD is stratified by *disease severity* (mild, moderate, and severe), *disease location* (upper gastrointestinal, ileal, ileocolonic, colonic, or perianal), *extent of disease*, and *disease phenotype* (e.g. penetrating, fibrostenotic, etc) [6].

MR helps to categorize the relative components of inflammatory, penetrating, or structuring disease in each individual patient, and it should be used for the surveillance of patients with known CD.

The assessment of the inflammatory activity is essential in order to delineate treatment strategies and determine the optimal choice and dose of medication; at the same time, the monitoring of patient is important to evaluate the efficacy of treatment [7].

In addition to dietary approach, several medical therapies are used to induce and/or maintain remission in CD, including corticosteroids, immunomodulators (e.g., azathioprine, mercaptopurine, methotrexate), and biologic agents (e.g., TNF-alpha inhibitors). In particular, with the recent advent of biologic therapies, follow-up to determine therapy efficacy and possible side-effects is becoming increasingly important.

Moreover, identification of location of fibrotic restricted lumen tracts and evaluation of prestenotic dilatation may help for operation planning, although this approach is foreseen after medical failure or complication occurrence [7]. However, when surgery is mandatory, bowel conservation is the principle aim.

In this context, MRI can demonstrate active small bowel inflammation recognizing wall thickening, ulcerations, increased wall enhancement, increased vascularity, perienteric inflammation, and reactive adenopathy. MR also allows more accurate identification of associated complications including penetrating and fibrostenotic disease as well as the extraintestinal manifestations that are usually associated with severe and long-standing intestinal inflammation [8–15].

However, it is important to keep in mind that, though they may be typical, radiological findings are not pathognomonic and can be detected in several diseases that fall within the differential diagnosis of CD [16, 17].

5.1 Wall Thickening

Although not pathognomonic, bowel wall thickening is the most reported finding in CD.

In adequately distended small bowel, the normal wall thickness should not exceed 3 mm [18].

It has also been shown that mural thickness increases with acute inflammation, generally ranging between 5 and 10 mm (given the associated histological findings of edema and inflammatory infiltrate) and that, in patients with CD, this well correlates with the presence and severity of the disease [19]. An associated thickening of the mucosal folds can also be detected (Fig. 5.1).

Even though the thickness decreases during remission, the inactive but yet pathological bowel is likely to be thicker compared with the healthy one [19, 20].

The most common site of CD is terminal ileum (Fig. 5.2), sometimes extending into the cecum. Discontinuous skip lesions may also be seen more proximally within the small and the large bowel (Fig. 5.3).

Mural thickness has been correlated to the degree of inflammation [21, 22].

The intestinal wall involvement may be symmetrical or asymmetrical, whereas greater involvement of the mesenteric border can lead to pseudosacculation.

Fibrosis and thinning of the diseased mesenteric wall lead to apparent dilatation of the opposing normal bowel wall (Fig. 5.4). Because all three bowel wall layers concur to form a real sacculation (in contrast to colonic diverticular disease), such a finding may also be referred to as a pseudosacculation [23]. Moreover, if there is suboptimal distention (as sometimes is the case, especially with jejunal loops), the disease may be over- or underestimated [18]. The evaluation of all images, sequentially acquired, may help to clarify the real extent of bowel involvement, as fluid distension in the small bowel will vary over time.

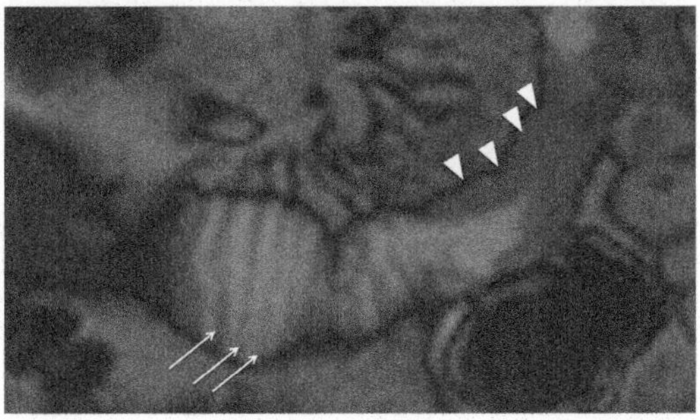

Fig. 5.1 Wall thickening. Magnified coronal True-FISP image shows wall thickening of a proximal ileal tract (arrowheads) and thickening of mucosal folds in the downstream segment (arrows)

Mural thickening can be appreciated on the whole sequences.

However, while thickness can be assessed with both T1- and T2-weighted sequences, balanced SSFP sequences should be avoided due to the "black border artifact," which can complicate the evaluation of bowel wall thickness. Conversely, HASTE sequence, which is insensitive to chemical shift and is not susceptible to black border artifact, is suitable for more accurate measurements [13, 18].

Fat-saturated T2-weighted sequences are very helpful for characterizing the etiology of wall thickening, appearing as an area of high signal intensity in the presence of active inflammation, due to the edema (Figs. 5.5 and 5.6); this finding has been shown a good correlation with the inflammation degree, and it can be particularly useful in patients with contraindications to contrast media administration [13].

In addition, in severe CD, the inflamed bowel wall has a layered appearance on T2-weighted sequences with a high signal intensity ring, representing the submucosal edema, between the mucosa and serosa ("target sign") [13, 24].

The hyperintensity of the submucosa also persists in the fat-saturated T2 images (Fig. 5.7).

Because on HASTE sequence, intramural fat deposition in chronic disease may result in elevated T2 signal intensity, fat-saturated T2-weighted sequences are also recommended to differentiate edema from fat: loss of signal on T2-weighted fat-saturated sequences is suggestive of chronic disease and this can be a helpful feature in interpreting the significance of thickened bowel wall [13, 24]. In these cases, a target appearance may be produced by low signal intensity "ring or halo" due to

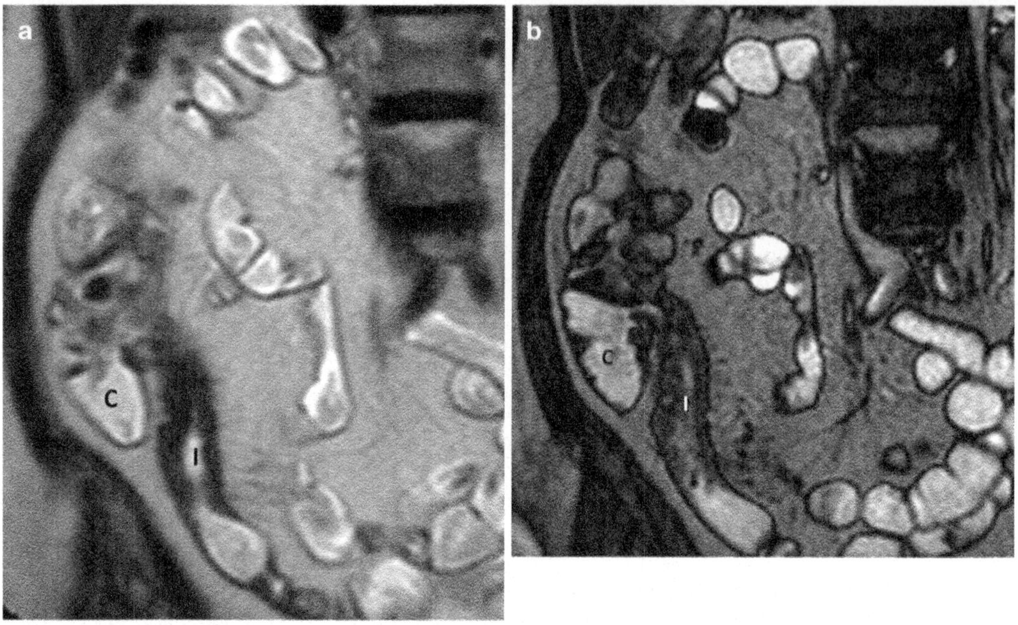

Fig. 5.2 Wall thickening. Magnified coronal HASTE (**a**) and True-FISP (**b**) images show wall thickening of the terminal ileum (I). C: cecum

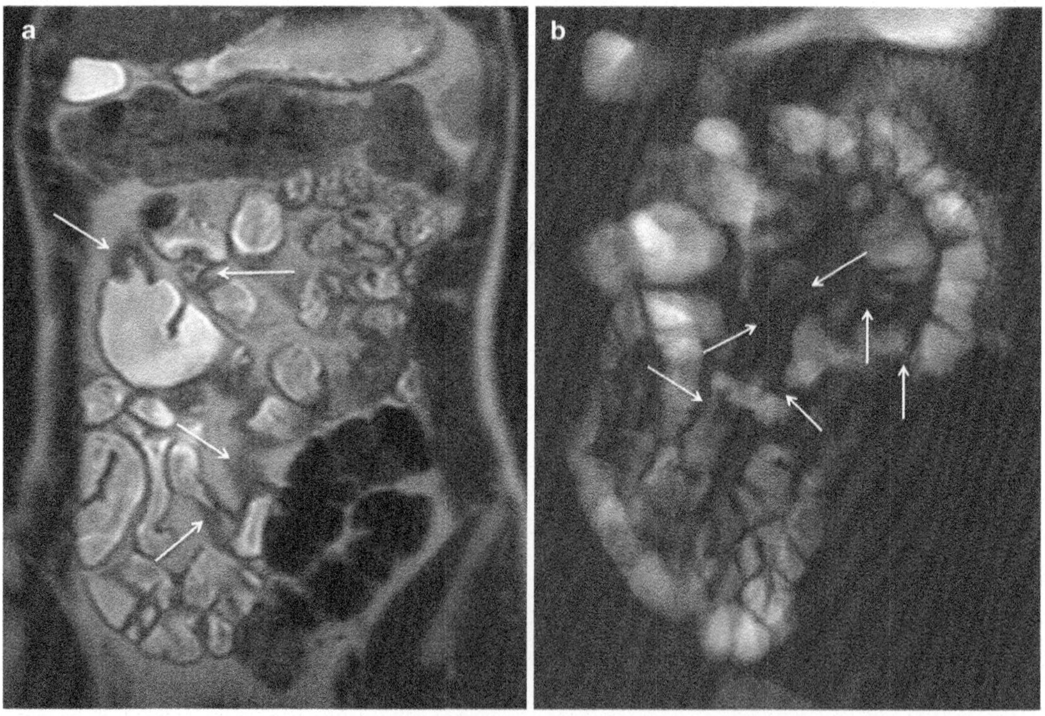

Fig. 5.3 Skip-lesions. Coronal T2-w HASTE (**a**) and coronal T2-w RARE thick-slab image (**b**) depict multiple inflammatory strictures (arrows) separated by segments of both normally distended and dilated small bowel segments

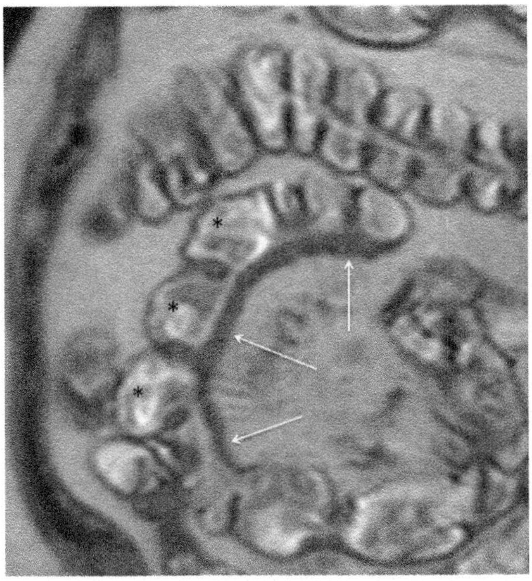

Fig. 5.4 Asymmetrical wall thickening and pseudosacculations. Magnified coronal T2-w HASTE image shows multiple pseudosacculations on the antimesenteric border (asterisks) due to asymmetric thickening of the bowel wall (arrows)

fat hypertrophy and fibrosis of the submucosa ("fat halo sign") (Fig. 5.8).

DWI, along with associated ADC measurements, has been shown to reflect abnormal activity in CD patients, helping to further characterize the bowel wall thickening and complementing the morphological information obtained by conventional MRI [25–27].

Particularly, active inflammatory thickening of the bowel wall is characterized by brighter signal in the b-value images and lower ADC values relative to normal segments (Fig. 5.9).

The visual assessment of DWI may provide higher accuracy, while calculation of the ADC may facilitate the quantitative analysis of the activity of the disease [25–27].

Nevertheless, a precise cut-off value for inflammation degree or for distinguishing inflammation from fibrosis has not been yet found.

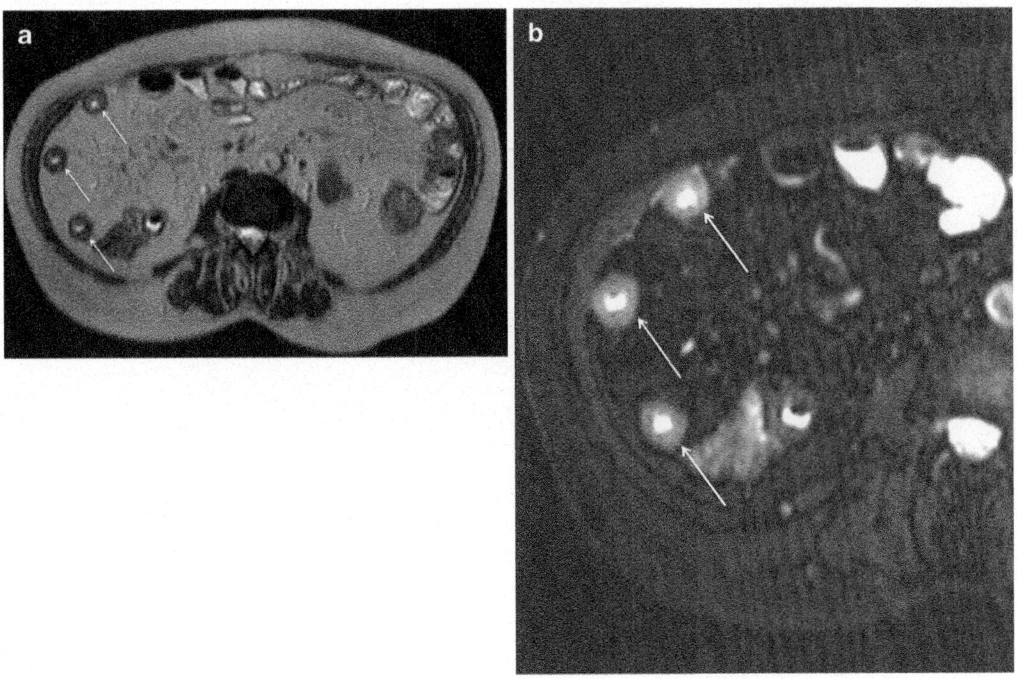

Fig. 5.5 Wall thickening in active Crohn's disease. Axial T2-w HASTE image (**a**) showing concentric bowel wall thickening in multiple segments (arrows). On magnified axial T2-w HASTE fat-saturated image (**b**), obtained at the same level, it is demonstrable mural hyperintensity of the diseased intestinal tracts, as a consistent finding of inflammatory edema in active Crohn's disease

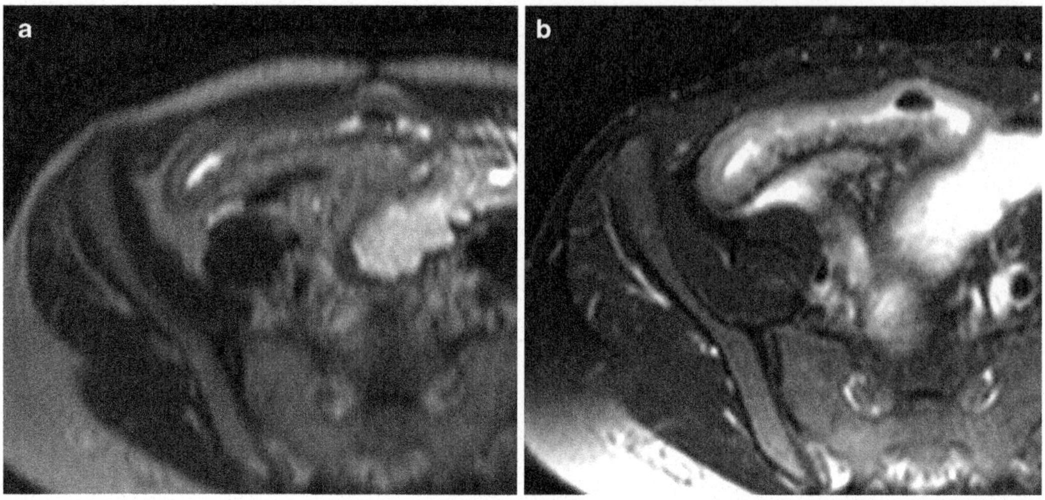

Fig. 5.6 Wall thickening in active Crohn's disease. Magnified axial T2-w HASTE image showing concentric wall thickening of distal ileum (**a**). Magnified axial T2-w HASTE fat-saturated image, obtained at the same level (**b**), depicts high signal intensity of the diseased intestinal segment, as a consequence of inflammatory edema, a consistent finding of active Crohn's disease

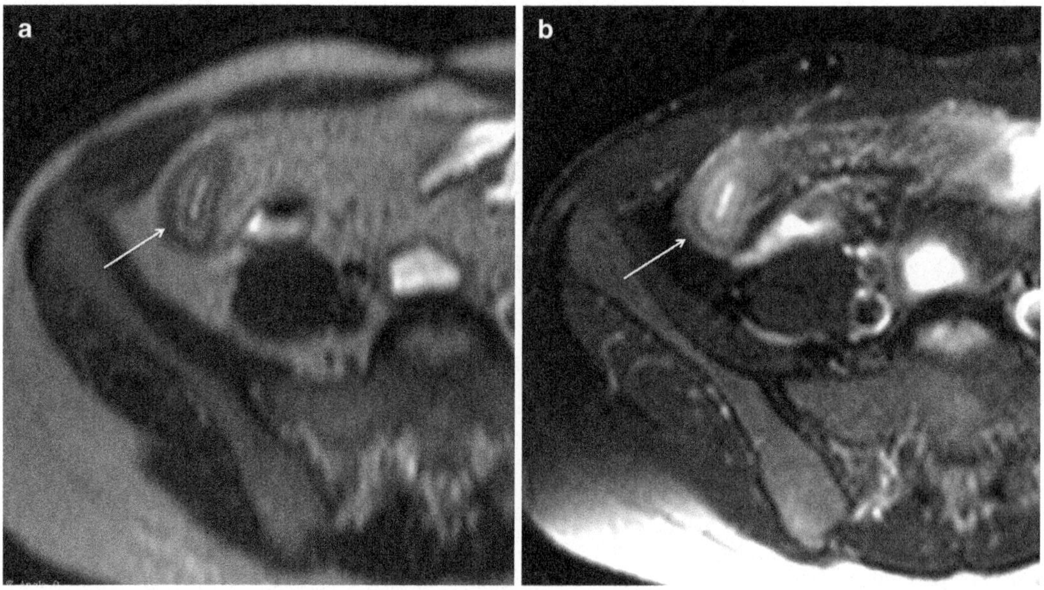

Fig. 5.7 *"Target sign"* in severe active Crohn's disease. Axial T2-w HASTE images in a patient with severe active Crohn's disease show a layered appearance of the inflamed terminal ileum (arrow), due to hyperintense submucosal edema (**a**), still visible employing fat-saturation (**b**)

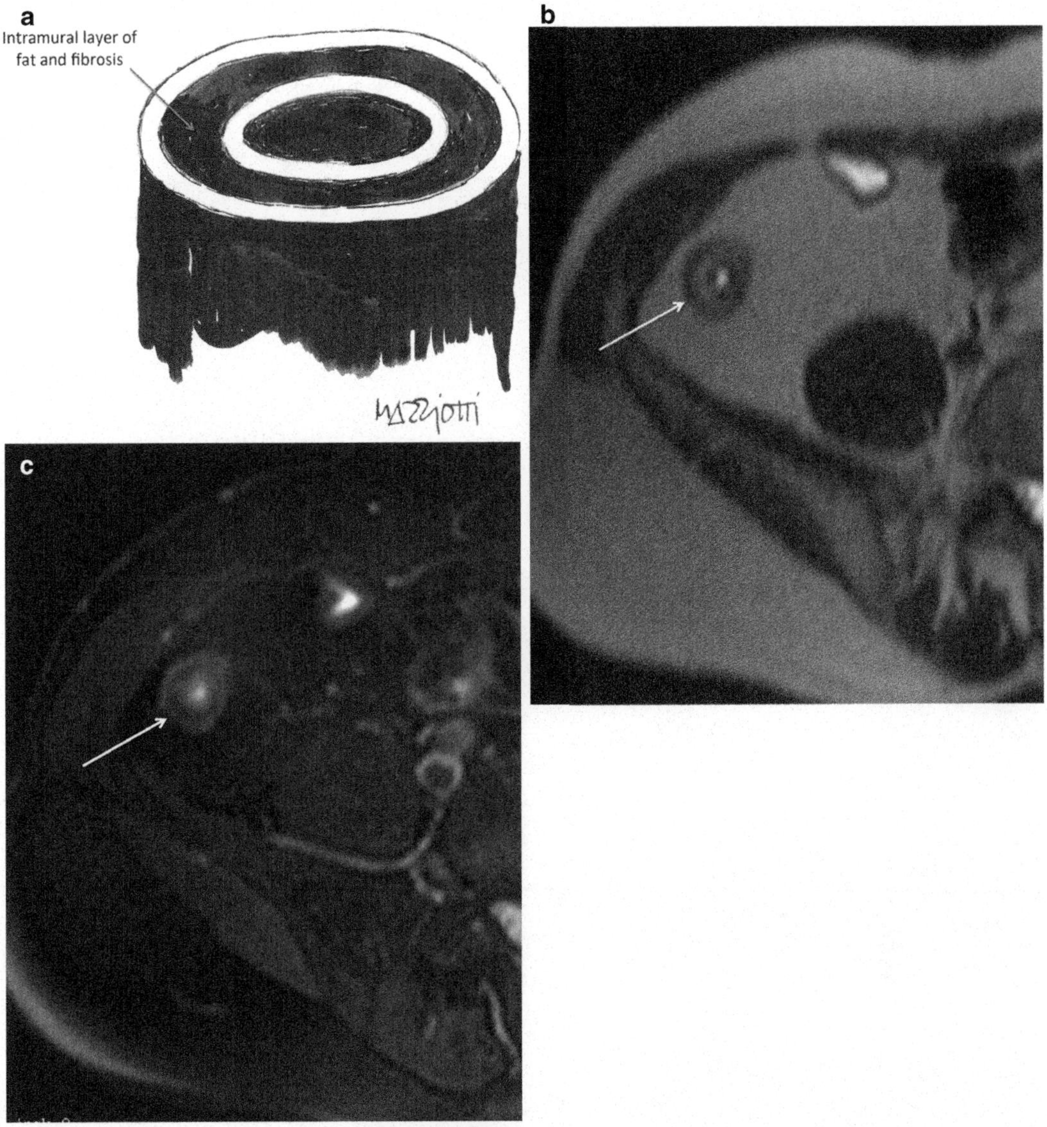

Fig. 5.8 *"Halo sign"* in chronic Crohn's disease. Illustration showing the "halo sign" (**a**), caused by fat hypertrophy and fibrosis of the intestinal submucosa in chronic Crohn's disease. Axial T2-w HASTE image (**b**) shows a layered appearance of the terminal ileum (arrow), which is due to hyperintense submucosal fibro-fatty proliferation, as confirmed by drop of signal intensity in the magnified T2-w HASTE fat-saturated image (**c**)

In presence of a borderline thickening of the small bowel wall, dynamic cine-MR sequences can more accurately demonstrate the number of pathological intestinal segments than what static MRI can do. In fact, abnormally decreased or increased peristalsis may be an early sign of the bowel involvement by CD and can potentially help to identify affected segments that would appear with only subtle or doubtful signs of inflammation on static images. However, it should be well kept in mind that additional factors over and above acute inflammation might affect motility in CD, including drug regimens, disease duration, location, coexistence of colitis, previous surgery, and mural fibrosis [28–31].

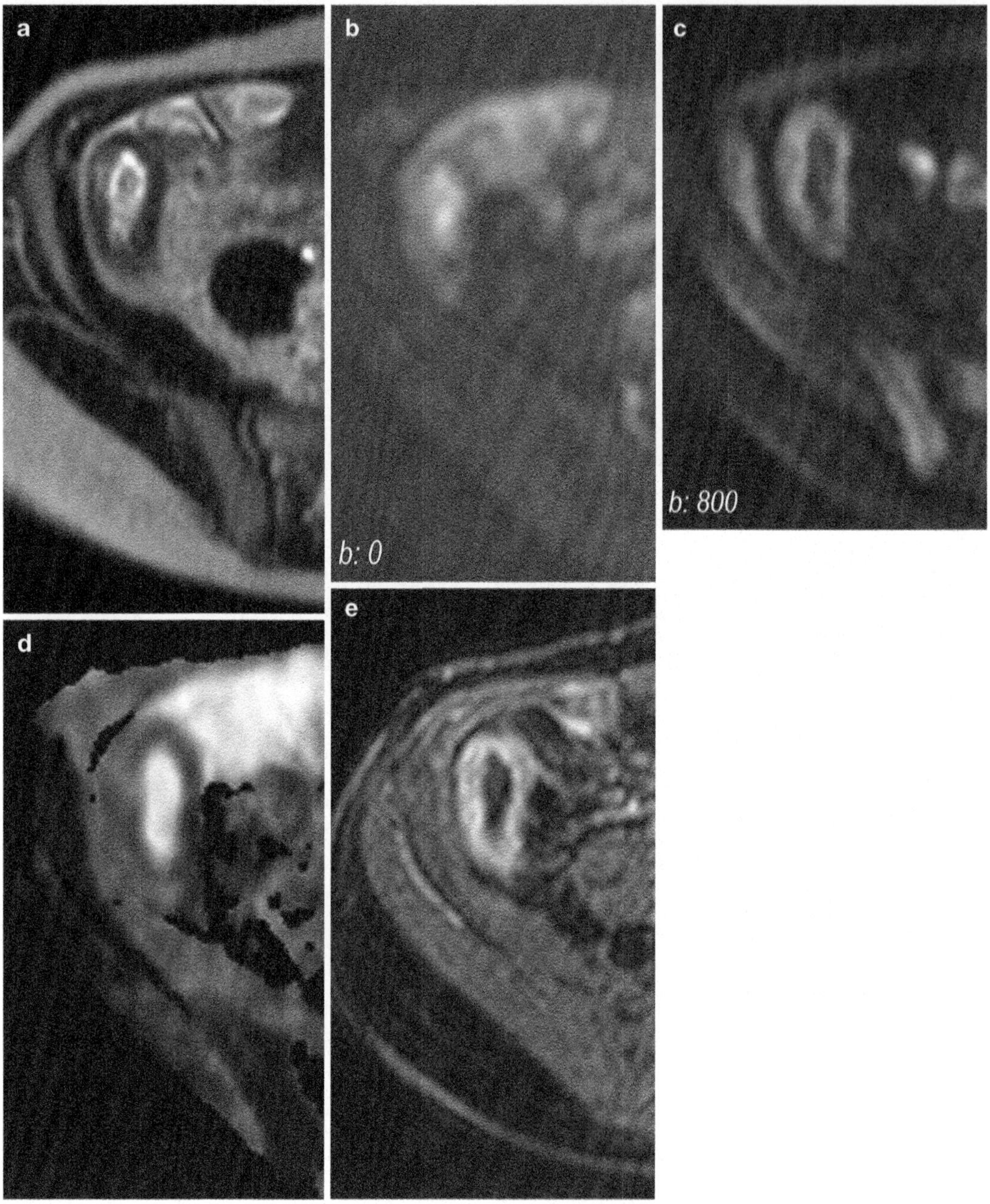

Fig. 5.9 Active Crohn's disease. Magnified axial T2-w HASTE image (**a**) shows wall thickening of the last ileal loop. Axial DW images obtained at different b-values (b: 0 s/mm^2 and 800 s/mm^2 respectively) show persistent hyperintense signal of the pathologically thickened wall of terminal ileum (**b**, **c**). The ADC map (**d**) demonstrates restricted diffusion (dark region) in the site of abnormal terminal ileum, which well correlates with marked hyperenhancement on axial fat-saturated T1-weighted image obtained after i.v. administration of gadolinium (**e**)

5.2 Ulcerations

The microscopic examination of the intestinal mucosa in patients affected by CD shows active inflammation, which is usually characterized by various degrees of injury of the epithelial crypts by neutrophils. In mild CD, only a small fraction of crypts is infiltrated by neutrophils, whereas in high degree of disease activity there is a corresponding escalation of the involved crypts and of the severity of crypt injury, which includes necrosis, abscess, and ulcer formation.

Two types of ulcers can be seen in CD: the superficial aphthoid ulcers and the deep fissuring ulcers, more problematic than the superficial ones. Detection of ulcerations is dependent on the quality of luminal distention.

The deep fissuring ulcers usually break through the mucosa and into the deeper layers of the bowel wall, causing initially submucosal inflammation and edema [12, 32, 33].

However, when the ulceration is initial and superficial, it is not always well demonstrable at MRI, even if full luminal distention is achieved; on the other hand, conventional fluoroscopy, if well performed, still holds this advantage.

Fissuring ulcers primarily manifest as areas of erosion in the mucosal lining, and they may finally extend into the submucosal space [12, 18, 32, 33].

Larger deep ulcers are outlined by luminal contrast material and appear as thin lines of high signal intensity longitudinally or transversely oriented into the thickened bowel wall (Fig.5.10).

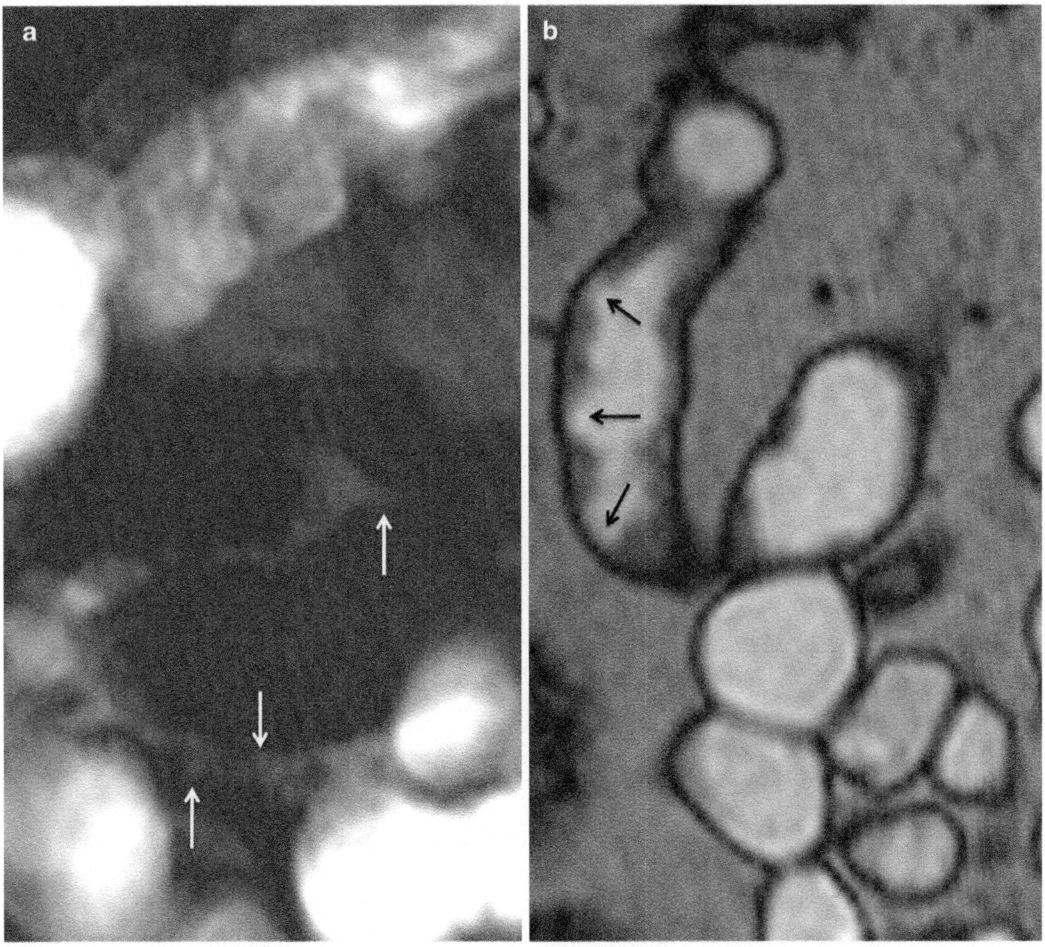

Fig. 5.10 Ulcerations. Magnified coronal T2-w RARE tick-slab (**a**) shows ileal narrowing and ulcerations (arrows). Coronal high-resolution True-FISP image (**b**) clearly depicts ulcerations as well (arrows)

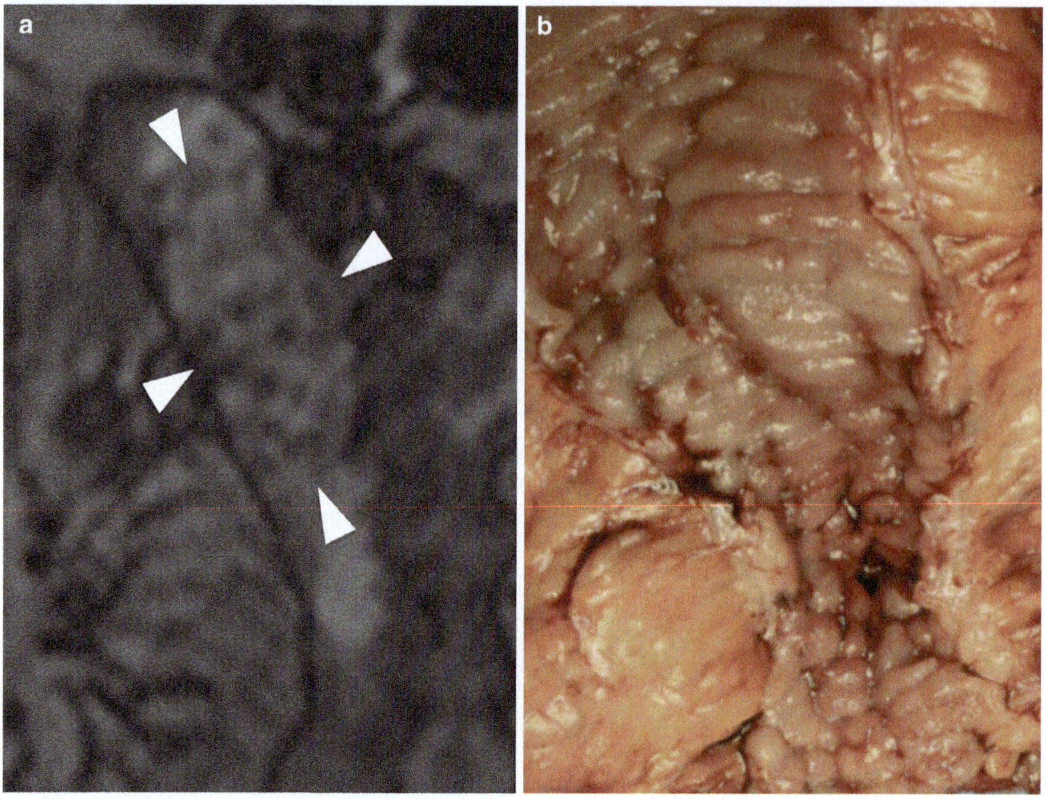

Fig. 5.11 *"Cobblestone appearance"* in Crohn's disease. Magnified coronal high-resolution true-FISP image (**a**) demonstrating the cobblestone appearance of edematous mucosa between longitudinal and transverse linear ulcers (arrowheads). Note the good correlation with the endoscopic findings (**b**)

HASTE and True-FISP images, which provide high contrast between the lumen and the bowel wall, may facilitate the detection of transmural ulcers [32].

The cobblestone aspect, which results from multiple confluent and intersecting longitudinal and transverse ulcerations with residual mucosal islands, appears as sharply demarcated and patchy areas of high signal intensity, a suggestive finding of advanced disease that may result in change of medical treatment (Fig. 5.11) [13, 32].

5.3 Increased Vascularity

The increased mesenteric vascularity adjacent to inflamed bowel loops is often present in the setting of acute inflammation.

In fact, when mesenteric blood flow increases, it results in engorgement of the vasa recta supplying the inflamed bowel segments, which appear as multiple tubular, tortuous structures on the mesenteric side of the ileum, aligned like the teeth of a comb ("comb sign") [13, 18, 32].

This finding, due to contrast enhancement of the vessels, is frequently best seen as high signal intensity parallel lines on contrast-enhanced T1-weighted fat-suppressed images (Fig. 5.12). Moreover, it can also be seen on fat-suppressed HASTE or True-FISP images, respectively, as high and low signal intensity parallel lines (Figs. 5.13 and 5.14).

The identification of comb sign, however, is much more limited on the nonfat-suppressed HASTE, due to the poor contrast resolution between vessels and mesenteric fat (Fig. 5.15).

The presence of the comb sign may suggest active disease [13, 18].

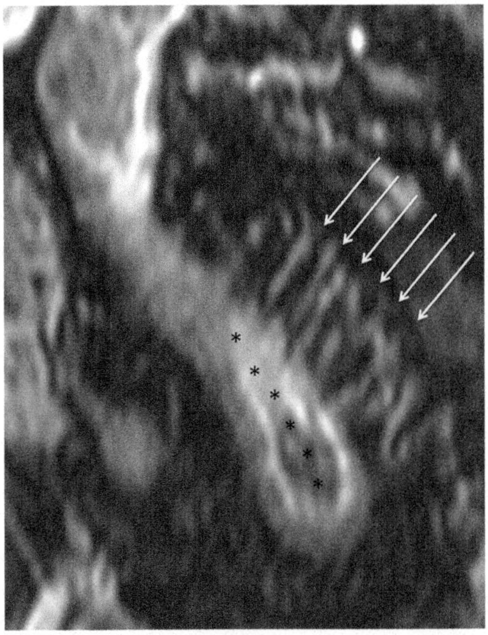

Fig. 5.12 *"Comb sign"* in Crohn's disease. Coronal GE T1-weighted and fat-saturated image, obtained after i.v. injection of gadolinium. Ectatic vasa recta (arrows) supplying a diseased bowel loop (asterisks)

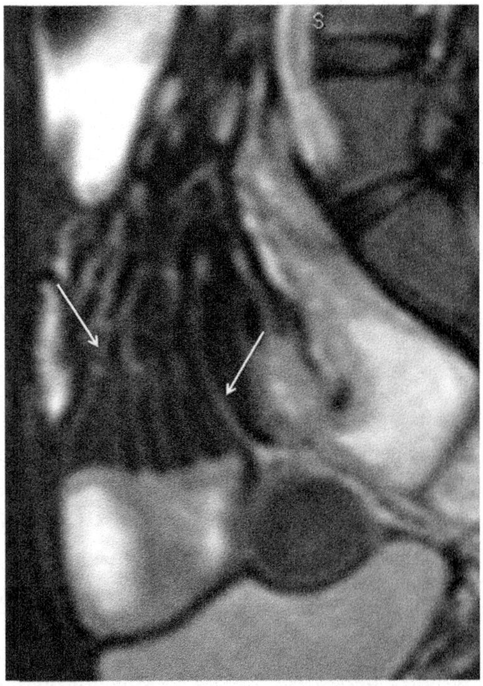

Fig. 5.13 *"Comb sign"* Crohn's disease. Sagittal T2-w HASTE fat-saturated image showing ectatic vasa recta as hyperintense linear structures (arrows), directed from mesentery toward a diseased bowel segment

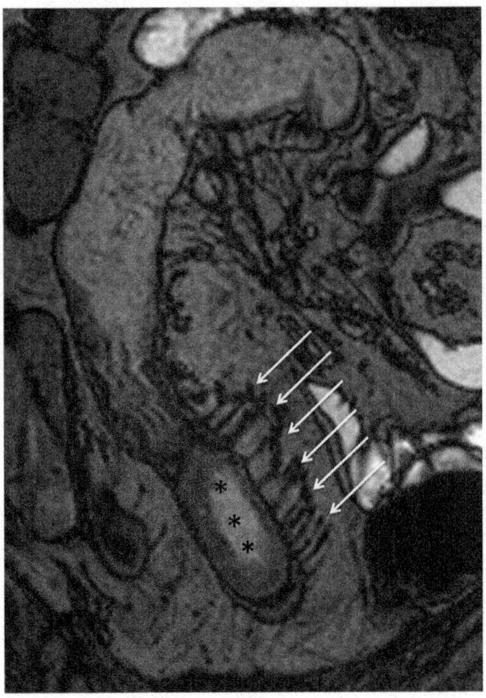

Fig. 5.14 *"Comb sign"* in Crohn's disease. Particular of coronal True-FISP image showing ectatic vasa recta, represented by multiple linear hypointensities directed toward the pathologic small-bowel tract (arrows)

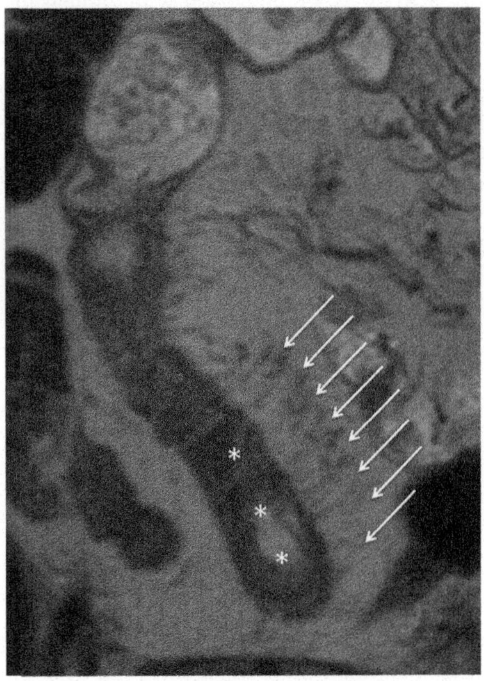

Fig. 5.15 *"Comb sign"* in Crohn's disease. Coronal T2-w HASTE image, without fat-saturation. The minor contrast between mesenteric fat and ectatic vasa recta does not allow an adequate visualization of comb sign

5.4 Patterns of Wall Enhancement

Bowel wall enhancement plays a very important role in determining disease severity, as it can be one of the earliest signs of inflammatory activity. It has been frequently reported to correlate with bowel inflammation and activity of the disease.

In CD, in fact, a marked increase in signal intensity of the inflamed bowel wall can be seen after intravenous gadolinium administration, due to increased tissue perfusion and vascular permeability.

The enhancement pattern of the inflamed bowel has also been studied to assess inflammatory activity as a comparison. It has been frequently reported to correlate with bowel inflammation and activity of the disease.

In CD, in fact, a marked increase in signal intensity of the inflamed bowel wall can be seen after intravenous gadolinium administration, due to increased tissue perfusion and vascular permeability.

The enhancement pattern of the inflamed bowel has also been studied to assess inflammatory activity as a comparison with clinical indices, and a significant enhancement decrease has already been correlated to a good response to medical treatment [18, 19].

The assessment of enhancement should be done during several phases, based on the different scanning times, relatively to contrast injection. Three main enhancement patterns can be described:

- asymmetric mural hyperenhancement, mainly represented at the mesenteric border,
- stratified mural hyperenhancement, visible at the inner or inner and outer side of the affected loops, due to edema, inflammatory infiltration, fat deposition or fibrosis of the submucosal layer,
- homogeneous symmetrical, with a transmural extension (Figs. 5.16, 5.17 and 5.18) [34].

The degree of the bowel wall enhancement has been proposed as a marker for evaluating the activity of the disease. However, estimating the intensity of enhancement remains largely subjective and can be done by comparing the abnormally enhancing segment to an adjacent normal loop or by likening bowel loops that are at similar

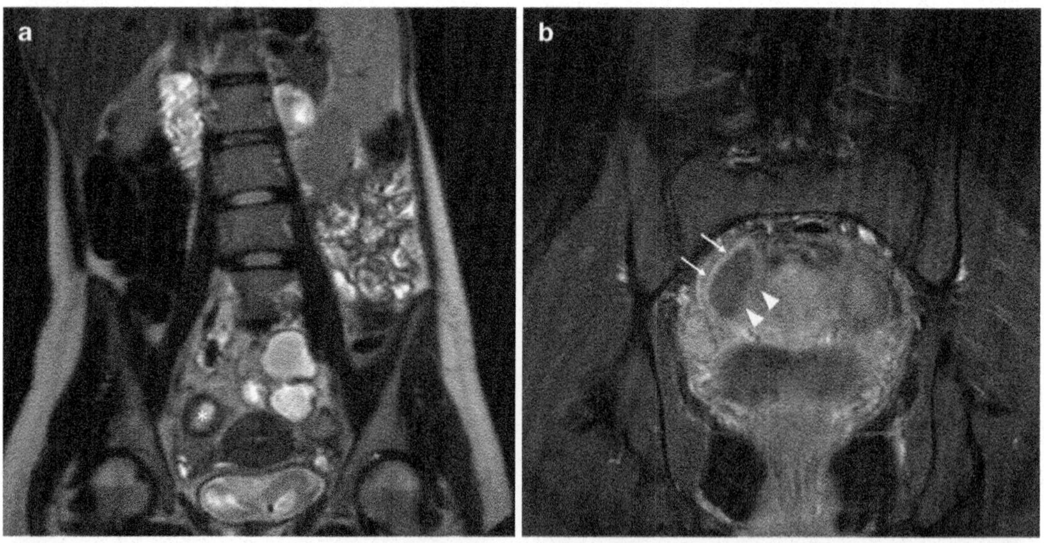

Fig. 5.16 On Coronal T2-w HASTE scan (**a**) wall thickening of an ileal loop is visible within the right lower abdominal quadrant (asterisk). Coronal T1-w scan with fat suppression image (**b**) demonstrates asymmetric mural enhancement, more pronounced on the right-cranial side (arrows) rather than the lower-left one (arrowheads)

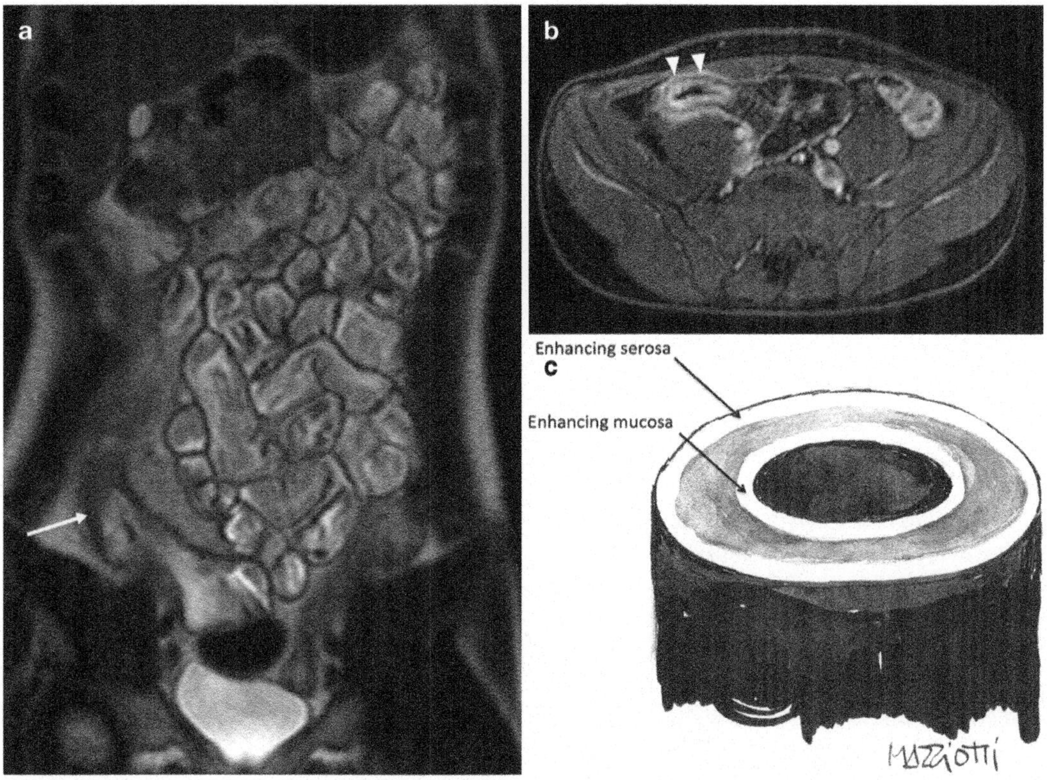

Fig. 5.17 Coronal T2-w HASTE scan (**a**) shows mural thickening of the last ileal loop (arrow) characterized by layered enhancement (arrowheads) on axial GE T1-w scan with fat suppression (**b**). Drawing illustrating the layered enhancement (**c**)

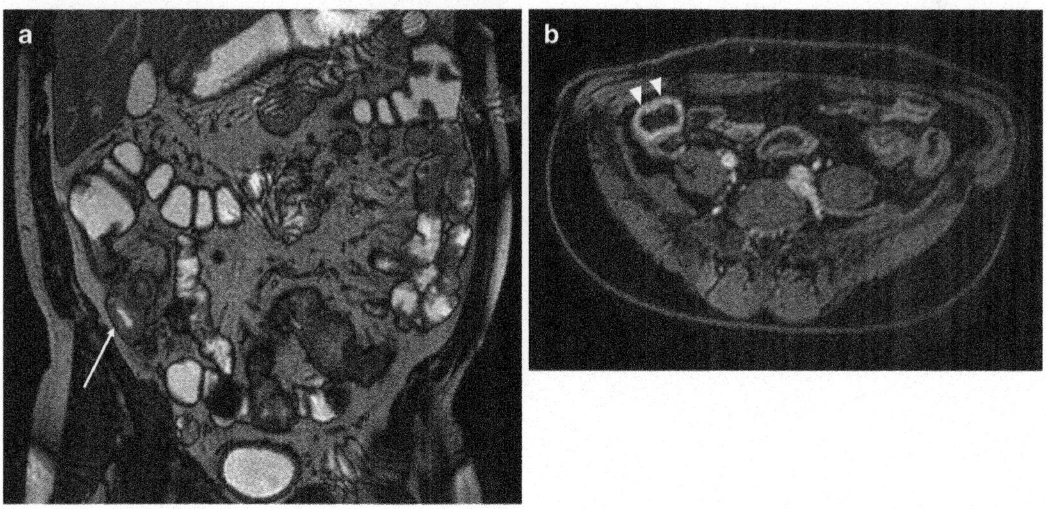

Fig. 5.18 Coronal True-FISP (**a**) demonstrating mural thickening of the last ileal loop (arrow). Homogeneous enhancement (arrowheads) is detectable on axial GE T1-w scan with fat-saturation (**b**)

distance from the center of the field of view; this is to mitigate the field inhomogeneity, which may otherwise influence the apparent level of enhancement.

The degree of enhancement has been determined by using subjective categorical scales (i.e., low, mild, high) or more objective grading methods, such as the comparison with renal or hepatic enhancement [35–38].

To quantify the degree of enhancement and thus providing more objective methods of measurement, several authors have calculated several enhancement ratios obtained by dividing postcontrast signal intensity by the baseline signal intensity. However, the use of such enhancement ratios may lead to errors, not least those associated with difficulty in intra- and interobserver placement of the region of interest (ROI), whose accuracy may be further limited in clinical practice. Moreover, there is often an overlap when assessing between active and chronic disease. It should also be taken into account that enhancement is evaluated as a snapshot in time using standard postcontrast sequences. On the contrary, the emerging dynamic enhancement techniques have been shown to be promising in quantifying enhancement, using this variable to determine inflammatory activity [7, 31, 39–41].

At DCE-MRI, the analysis of the time-dependent changes of signal intensity after gadolinium administration adds valuable information about the activity of Crohn's disease, as well as the kinetic of the signal variation reflects the status of tissue microcirculation, allowing semi-quantitative (slope of enhancement) and quantitative (intravascular-extracellular transfer coefficient, volume of extracellular fluid) measurements, which may ultimately prove to be more useful in determining the level of inflammatory activity [7, 31, 39–41].

In addition, it should always be kept in mind that inadequate bowel loop distention can affect the assessment of enhancement: nondistended loops can brightly enhance, a finding that may inadvertently be interpreted as active disease.

5.5 Perienteric Inflammation

Edema and fluid in the mesenteric fat of an involved bowel segment are present in most of patients with advanced inflammatory disease, and they usually are proportionate with the activity of the disease [18].

Edema tracks along the adjacent mesentery from an inflamed bowel loop, and it is more evident on the T2-weighted fat-suppressed images [13]. The increased enhancement of the mesenteric fat around a bowel segment is another secondary sign of perienteric inflammation (Fig. 5.19).

Mesenteric inflammation may also lead to the formation of adhesions that can cause bowel kinking and obstruction. They may not be directly visualized, but their presence can be inferred on the basis of kinking, tenting, or obstruction of the bowel segments in the absence of another associated abnormality [32]. An early detection of adhesions is important because they may require a treatment if they cause obstruction.

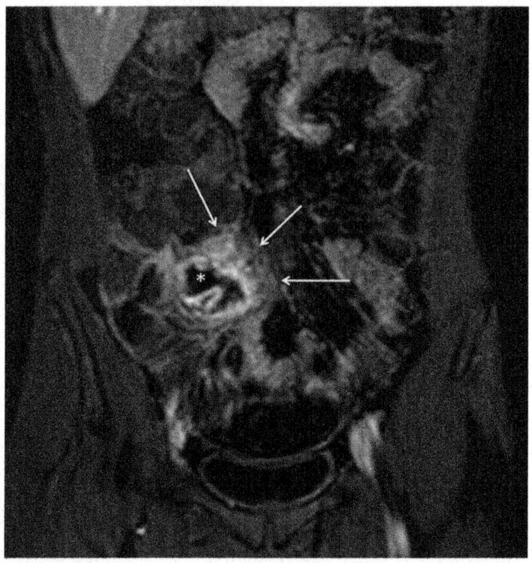

Fig. 5.19 Perienteric inflammation. Coronal T1-w fat-saturated GE image obtained after i.v. administration of gadolinium shows enhancement of both the wall of terminal ileum (asterisk) and of perivisceral fat (arrows), as a consequence of progression of the inflammatory process to the adjacent mesentery

Thicker, fibrous adhesive bands may appear as isointense structures with enhancement similar to that of sinuses or fistulas. However, adhesions are best differentiated from sinuses and fistulas by means of MR fluoroscopy, because kinking and stretching of bowel loops may be better observed in real time.

Furthermore, MR fluoroscopy is also very helpful for grading the severity of bowel obstruction.

5.6 Reactive Lymph Nodes

Reactive nodes and adenopathies are common in CD patients and are defined by a short axis greater than 1–1.5 cm [1, 8, 9, 13, 18, 19, 24, 31–35].

They are usually detectable at MRI and typically appreciable in proximity to the pathologic segment.

Due to the poor contrast resolution of the extraenteric mesenteric fat tissue, lymph nodes are more easily identified on T2-weighted fat-suppressed or True-FISP images rather than on HASTE sequences (Figs. 5.20 and 5.21) [18].

Lymph nodes that are increased in size can also be seen, both in active and inactive CD disease, and their presence alone cannot be considered as an inflammatory activity marker. However, nodal edema, visible at fat-saturated HASTE acquisitions, and homogeneous contrast enhancement, even in normal-sized nodes, is highly suggestive of active inflammation in CD patients (Fig. 5.22) [42, 43].

Although controversies exist in literature about considering number and size of lymph nodes as inflammatory markers, enhancement characteristics and DWI parameters tend to be similar to those of the affected bowel loop in CD patients [44–46].

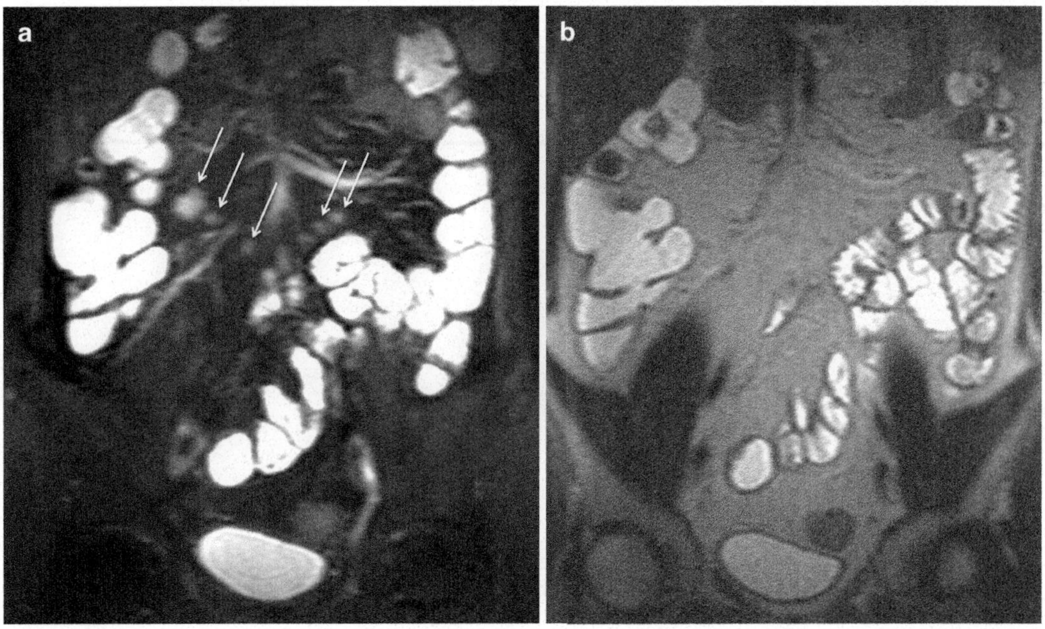

Fig. 5.20 Reactive adenopathy. Coronal T2-weighted fat-saturated image (**a**) well depicts multiple mesenteric lymphadenopathies (arrows), less evident on coronal T2-weighted image, performed at the same level without fat-saturation (**b**)

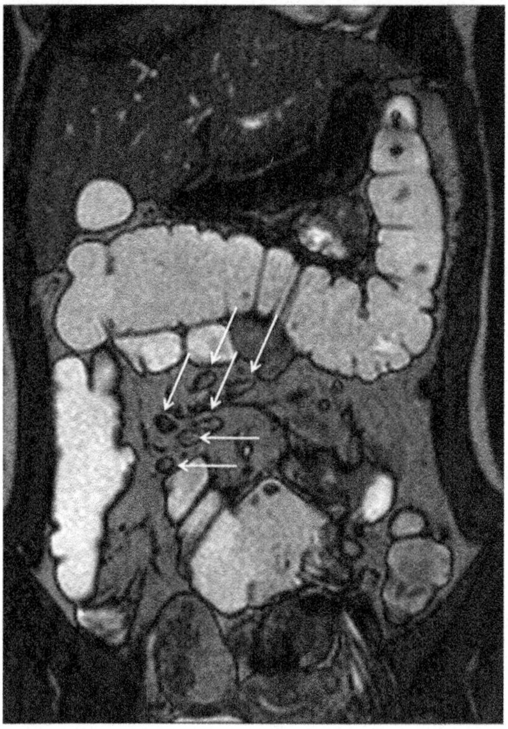

Fig. 5.21 Reactive adenopathy. Coronal True-FISP image shows multiple mesenteric lymph nodes (arrows)

5.7 Mesenteric Fibrofatty Proliferation

Abnormalities of the mesenteric fat tissue, including its hypertrophy and the fat-wrapping, have been long recognized on surgical specimens and can be defined as the increased mesenteric fat tissue that produces a mass effect with separation of bowel loops (Figs. 5.23 and 5.24) [7, 31, 39–41].

Frequently, it is also asymmetric, involving preferentially the mesenteric border of the bowel, even if fibro-fatty proliferation often completely encircles the involved bowel loops.

Fat-wrapping usually occurs in patients with long-standing, established transmural inflammation, and it is a specific feature of chronic CD [13].

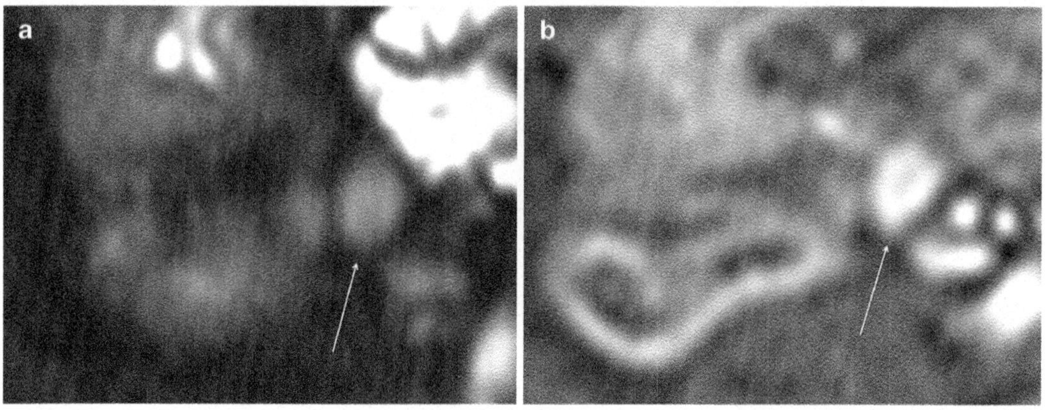

Fig. 5.22 Reactive adenopathy in active Crohn's disease. Magnified coronal T2-w fat-saturated HASTE image (**a**) shows an enlarged and hyperintense lymph node (arrow) that avidly enhances after i.v. administration of gadolinium, on axial GE T1-weigthed image (**b**)

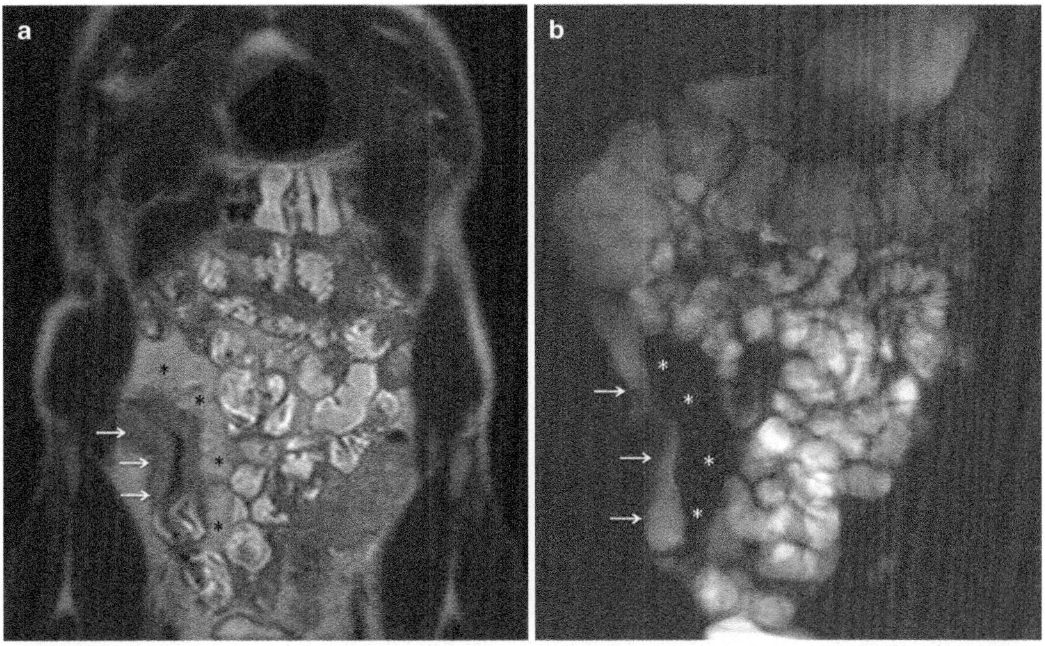

Fig. 5.23 Fibro-fatty proliferation in chronic CD. Coronal T2-w HASTE image (**a**) shows the displacement of the normal small bowel loops away from the diseased ileal segment (arrows), as a consequence of mesenteric inflammation and fibro-fatty proliferation (asterisks). Coronal T2-w RARE thick-slab image (**b**) clearly demonstrates the luminal narrowing of the pathologic ileal loop (arrows)

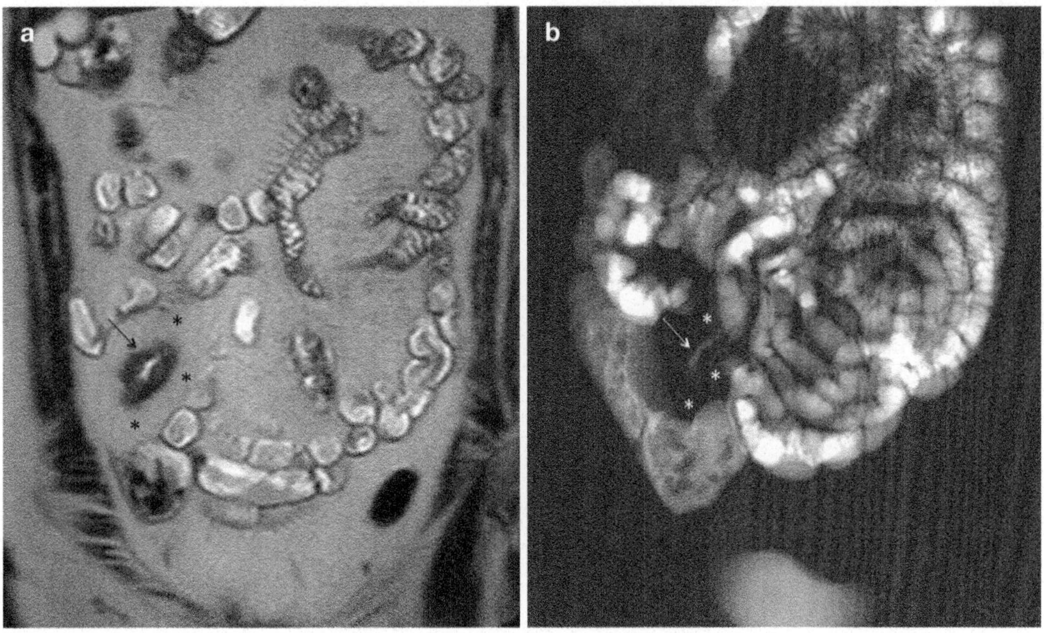

Fig. 5.24 Fibro-fatty proliferation in chronic CD. Coronal T2-w HASTE scan (**a**) shows the normal small bowel displacement away from the diseased ileal loop (black arrow), caused by mesenteric inflammation and fibro-fatty proliferation (asterisks). Coronal T2-w RARE thick-slab image (**b**) also shows lumen narrowing of terminal ileum (white arrow)

5.8 Penetrating and Stricturing Patterns in CD

Crohn's disease is a heterogeneous entity. Clinically, the disease is classified into three categories, which are determined by pathological manifestations and symptoms, site of disease in the gastrointestinal tract, and the presence of complications, such as strictures or penetrating disease.

As previously stated in the introductive part of MRI findings regarding the phenotype of the disease, in the classification of Vienna, the behavior of Crohn's disease was separated into three prognostic relevant entities, including nonstricturing and nonpenetrating disease (B1), stricturing disease (B2), and penetrating disease (B3) [47]. This classification system was later modified in the classification of Montreal, by adding perianal penetrating disease, because perianal fistulas and abscesses have a different prognosis and outcome than intra-abdominal penetrating forms of CD [48]. Although these classification systems were originally designed to CD patients for clinical studies, nowadays, the behavior of the disease can also be used to determine the most correct therapeutic decision. In fact, for what concerns the nonstricturing/nonpenetrating (or inflammatory) phenotype, it is usually medically treated.

Conversely, the presence of stricturing and penetrating disease influences the decision on surgical intervention and its timing.

The penetrating phenotype would eventually require surgical treatment in several cases. The stricturing phenotype requires intervention with mechanical treatments, such as balloon dilatation, strictureplasty, or resection, in many cases.

MR of the small bowel can offer a similar phenotypic characterization of Crohn's disease, based on the characteristic imaging findings, differentiating penetrating and structuring disease from the inflammatory phenotype [24, 49].

It is important, therefore, to report about the presence of penetrating ulcers, sinus tracts, fistulas, inflammatory masses, abscesses, and strictures/stenosis while interpreting MR enterography examinations.

5.8.1 Penetrating Disease

The fistulizing/penetrating phenotype is characterized by severe inflammation, which causes the progression of the transmural ulcerations into the surrounding mesentery, resulting in sinus tracts or fistulas.

Fistulas occur in up to one-third of CD patients, a quarter of which are enteroenteric. The lifetime risk ranges from 20% to 40%. The reported sensitivity of MR to detect internal fistulas ranges from 83.3% to 84.4%, while its specificity is approximately 100% [8, 12, 50, 51].

A sinus tract is defined as a blind-ending tract that arises from the bowel wall but does not reach another epithelium-lined surface. Conversely, fistulas may bridge adjacent loops of the small bowel or cross from the small bowel to the colon, stomach, and any adjacent organs, as well as the skin. Internal fistulas are more common than external enterocutaneous fistulas, and enteroenteric fistulas are usually asymptomatic [13, 24, 49].

On images, sinus tracts appear as nodular irregularities and spiculations adjacent to serosal surface of the bowel. Small sinus tracts are usually difficult to identify, because of partial volume averaging, but thin-section MR images can be useful in assessing such tracts [24]. In particular, images obtained along a plane perpendicular to the bowel segment allow more accurate assessment of inflammatory perivisceral changes and of the transmural inflammation.

Large sinus tracts may be outlined by enteral contrast material and are seen on T2-weighted sequences as high signal intensity linear features.

Fistulas are usually well visualized on the True-FISP and HASTE sequences, due to high signal of their fluid content; however, they can also be seen as rim-enhancing low-signal tracts on contrast-enhanced T1-weighted fat-suppressed images [13, 24, 49]. Additionally, the multiplanar imaging capability of MR can be exploited to better depict complex fistula tracks.

When fistulas are complex, they appear as a network of intersecting linear tracts (Figs. 5.25 and 5.26). Very often in these cases, concomitant small sinus tracts can be depicted.

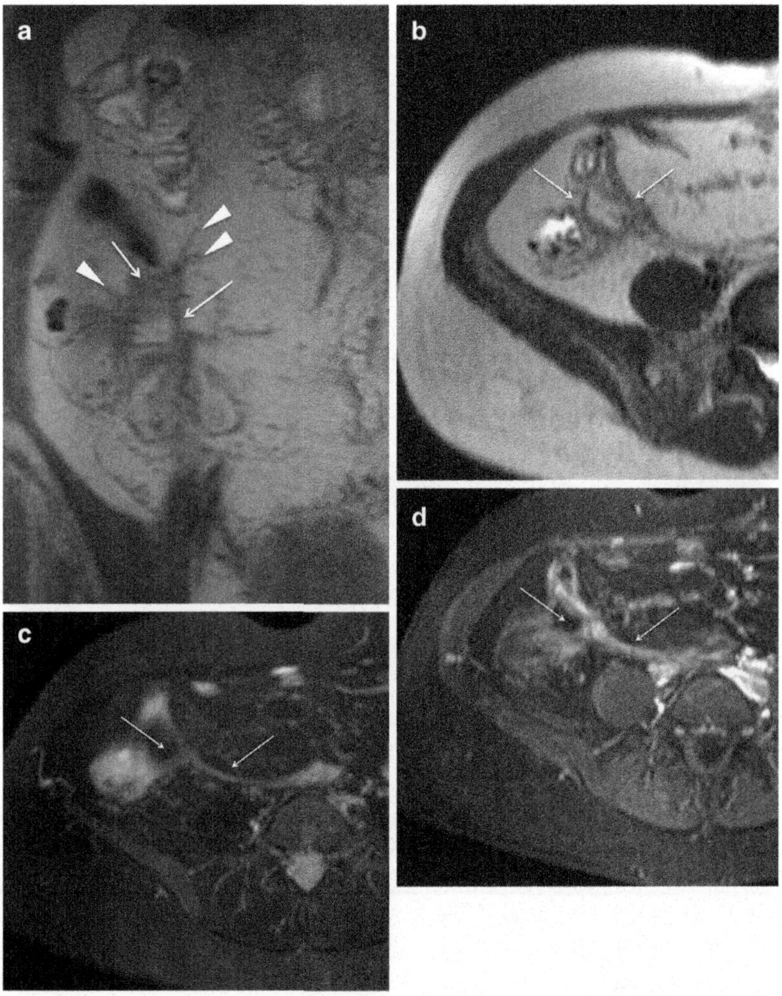

Fig. 5.25 Enterocolic fistula. Magnified coronal (**a**) and axial (**b**) T2-w HASTE acquisitions showing enterocolic fistulas (arrows). The route of the fistula (arrows) is well detectable on axial T2-weighted fat-saturated image (**c**) and axial T1-w fat-saturated GE image obtained after i.v. administration of gadolinium (**d**). Note also the coexistence of a small sinus tract (arrowheads) in (**a**)

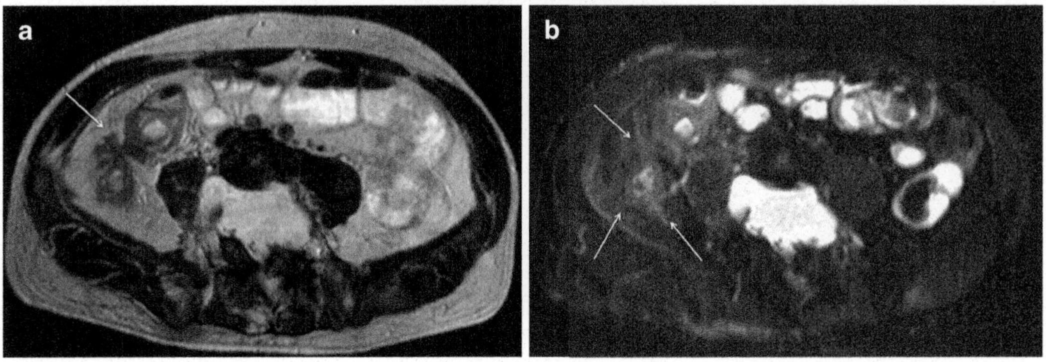

Fig. 5.26 Enterocolic fistula. Axial T2-w HASTE image (**a**) showing enterocolic fistulas (arrow). Perivisceral edema is more evident on axial T2-w fat-saturated HASTE image (**b**)

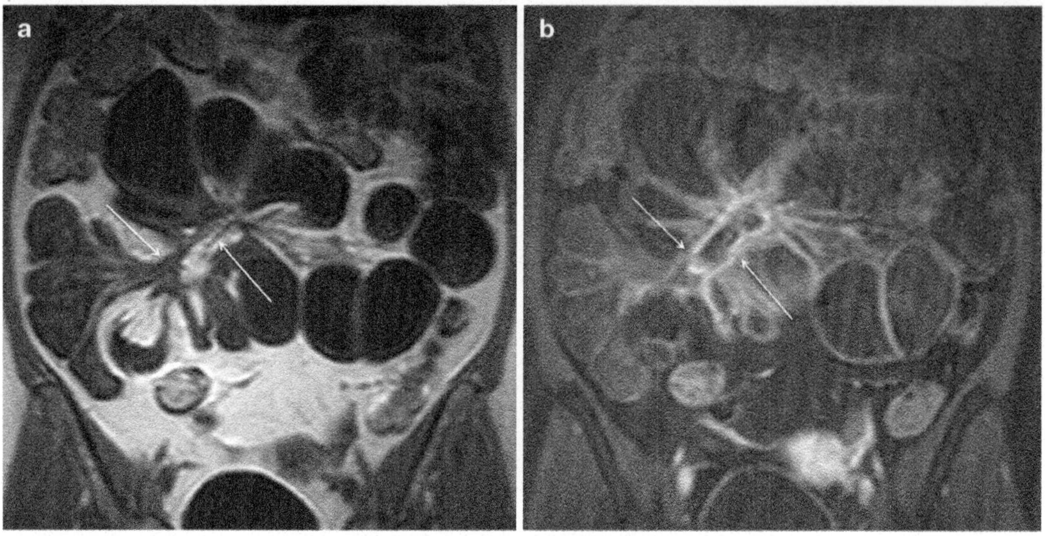

Fig. 5.27 Enteroenteral fistula. Coronal T1-w GE image (**a**) showing a "stellate pattern" caused by bowel loop retraction to a central area (arrows), with an associated complex enteroenteral fistula. Coronal T1-w fat-saturated GE image, obtained after i.v. injection of gadolinium, shows inflammatory hyperenhancement and "star sign" (**b**)

Typically, a complex network between multiple closely adherent and converging loops of an inflamed bowel tract may appear as a stellate configuration on contrast-enhanced MR images, the "star sign," a suggestive finding of enteroenteric fistulas (Fig. 5.27) [13].

Fistulas, if associated with acute inflammation, manifest themselves as conspicuous bright signal on DWI sequences. Both contrast-enhanced and DWI sequences have been shown to equally improve detection of fistulas when combined with a T2-weighted sequence; DWI may be especially useful in patients with risk factors for contrast agents [26, 27]. Intestinal fistulas by themselves are not a primary indication for surgery. In fact, surgery is indicated if fistulas involve the renal tract causing renal impairment or infection, if their drainage may cause personal discomfiture and hygiene, or if they create a significant bypass resulting in intestinal malabsorption.

Penetrating disease can also determine the formation of phlegmons or abscesses.

In fact, if deep transmural ulcers eventually penetrate the bowel muscle layers, they can cause inflammation in the adjacent mesenteric tissue, which may lead to the formation of phlegmons evolving in peri-intestinal abscesses. Moreover, deep ulcers can also form cavities within the bowel wall, which may become secondarily infected. Mural abscesses, causing bulging of the bowel wall, may also perforate it leading to the formation of intra-abdominal abscesses, just in the adjacent mesentery [13, 24, 49]. Phlegmons appear as heterogeneous "pseudomass lesions," with low signal on T1-weighted images and intermediate to high signal on T2-weighted images.

Abscesses appear as fluid collections that show intense peripheral enhancement, with or without associated intralesional air (Figs. 5.28, 5.29, 5.30 and 5.31) [13]. However, interloop abscesses may be difficult to detect when bowel loops are retracted or matted.

On DWI, abscesses appear bright and their conspicuity on this sequence can enhance the detection and diagnostic confidence (Fig. 5.31) [26, 27].

A correct diagnosis is important, because the absence of an abscess in the presence of a penetrating disease often alters medical therapy; in fact, the use of steroids is usually avoided in such cases.

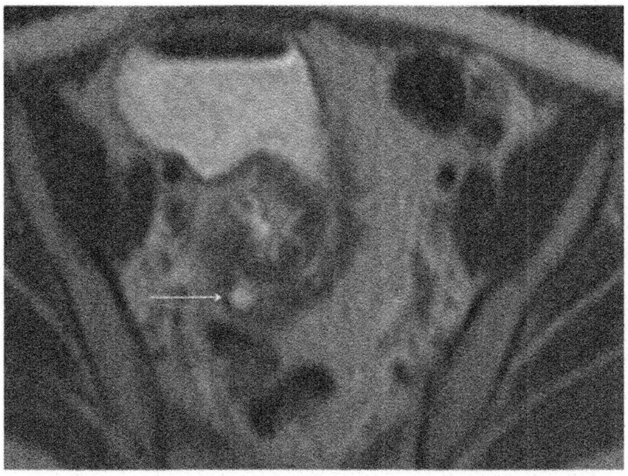

Fig. 5.28 Intramural abscess. Axial T2-w HASTE scan showing a round hyperintense lesion (arrow) within the thickened wall of a diseased ileal segment. Of particular note, the lumen narrowing and the distention of prestenotic tract

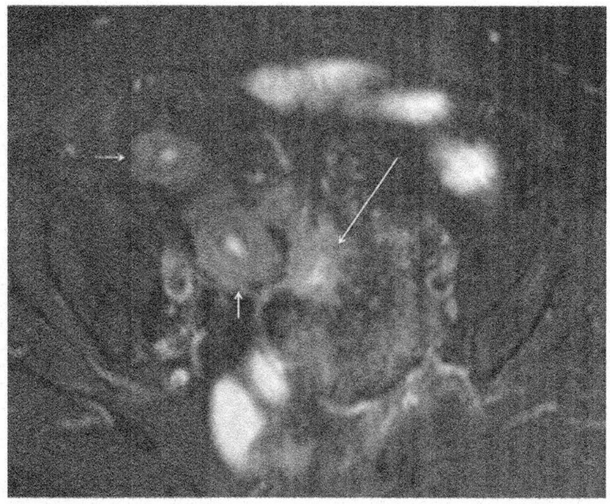

Fig. 5.29 Perivisceral phlegmon evolving toward an abscess. Axial T2-w fat-saturated HASTE image shows a mesenteric pseudomass (long arrow) adjacent to diseased bowel segments (short arrows) and with a small central colliquative area

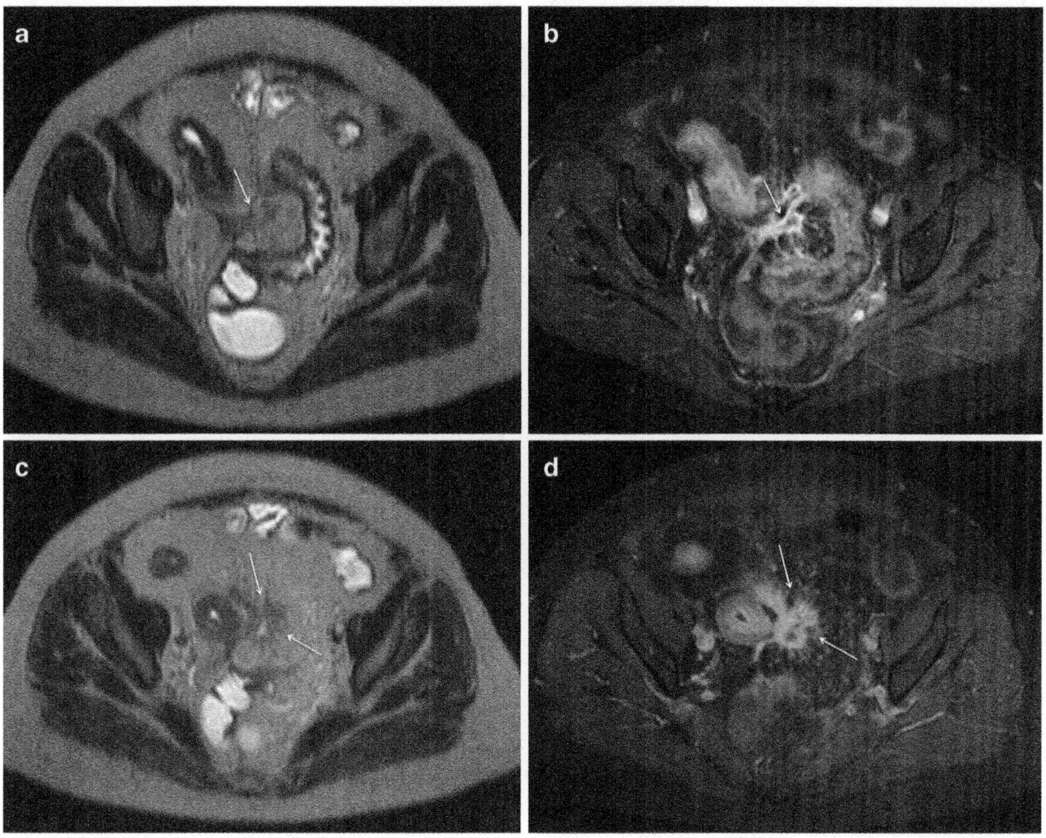

Fig. 5.30 Enterosigmoid fistula with perivisceral abscess. Axial T2-w HASTE images (**a**, **c**) and axial T1-w fat-saturated GE acquisition, performed after i.v. injection of gadolinium (**b**, **d**) obtained along two different axial abdominal planes. (**a**) and (**b**) show an enterosigmoid fistula (arrow). (**c**) and (**d**) depict a small abscess in the adjacent mesentery, just below the fistula (arrows). Note also the "target sign" in the adjacent diseased bowel segment (**c**, **d**)

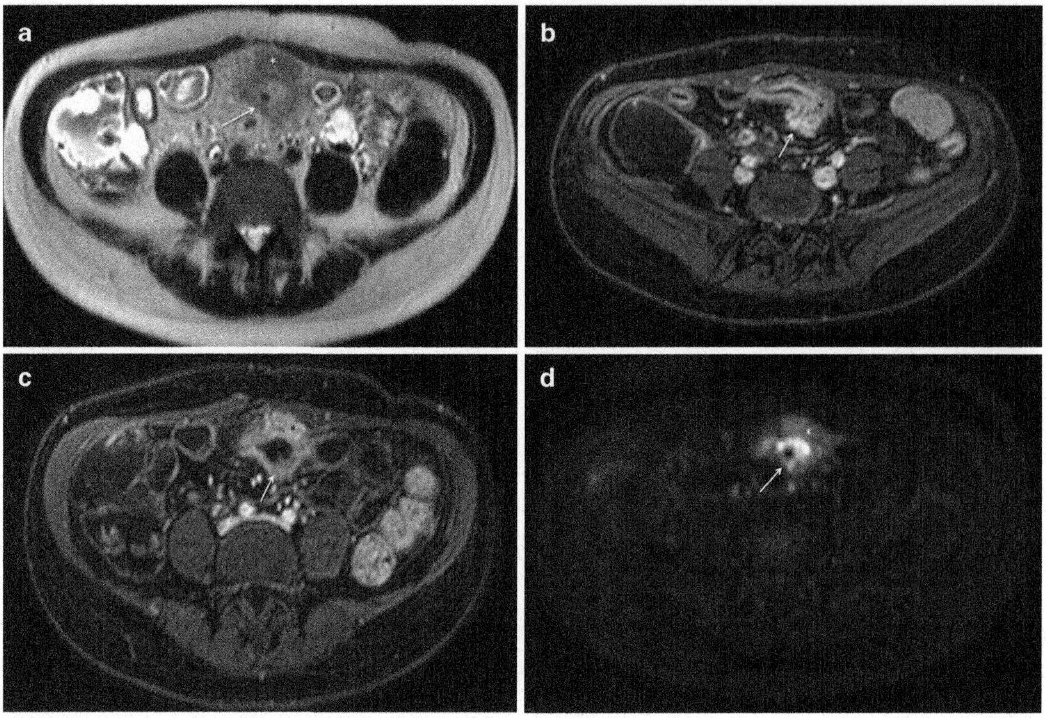

Fig. 5.31 Perivisceral abscess with gas bubble. Axial T2-w HASTE scan (**a**) demonstrating a mesenteric abscess with a small gas bubble inside (arrow) adjacent to a diseased bowel segment (asterisk). Axial T1-w fat-saturated GE images, obtained after i.v. injection of gadolinium at different levels (**b**, **c**), demonstrate diffuse enhancement of inflamed bowel segment (asterisk), which appears to be undetachable from the enhancing abscess (arrow). On axial DW image performed at b: 800 s/mm^2 (**d**) abscess appears hyperintense (arrow), as well as the diseased bowel tract (asterisk). Note the signal void of the gas bubble

5.8.2 Fibrostenosing Disease

Over time, chronic inflammation within the bowel wall progresses to mural fibrosis. Additionally, when fibrosis is associated with stricture formation, bowel wall obstruction may develop.

A small bowel stricture is defined as a luminal narrowing (at least 50% less than the healthy bowel loops) usually but not necessarily associated to an upstream luminal dilation.

The latter can be classified as mild if the luminal caliber is comprised between 3 and 4 cm or moderate to severe if it exceeds 4 cm.

The importance of identifying fibrotic strictures with certainty is due to the fact that they are unresponsive to medical therapy [18, 34].

Commonly, a single stricture in the terminal ileum causes the obstruction, and treatment is carried out by resection and primary anastomosis. However, recurrent disease after ileocolic resection is seen in one-third of patients within 10 years.

Moreover, the presence of a fibrotic stricture does not exclude the possibility of coexistent active inflammation elsewhere in the bowel [24, 49]. One of the other more common procedures is strictureplasty, which is used to open up a narrowed section of the bowel.

Unlike a resection surgery, no part of the bowel is removed during a strictureplasty, making it an appealing alternative to a resection when possible. Strictureplasty may be done alone, or it may be done at the same time as a resection.

MRI can provide useful information in these settings by differentiating between fibrotic and inflammatory strictures. In the fibrostenosing disease phenotype, MRI demonstrates a fixed narrowing of the involved bowel with wall thickening associated and with a marked prestenotic luminal distention. The thickened submucosa of a strictured, fibrotic bowel segment, in the absence of active disease, does not typically show any increased signal intensity on T2-weighted images, because of the lack of edema [13, 20].

On MR-fluoroscopy and cine images, fibrotic strictures appear as aperistaltic bowel segments that often display a fixed mural thickening and luminal narrowing.

Progressive enhancement gain in delayed contrast phases and low ADC values derived from DWI scans are suggestive of fibrosis (Fig. 5.32) [22, 52, 53].

On the other hand, if an early uptake of gadolinium is suggestive of active inflammation at perfusion MRE, this technique has demonstrated

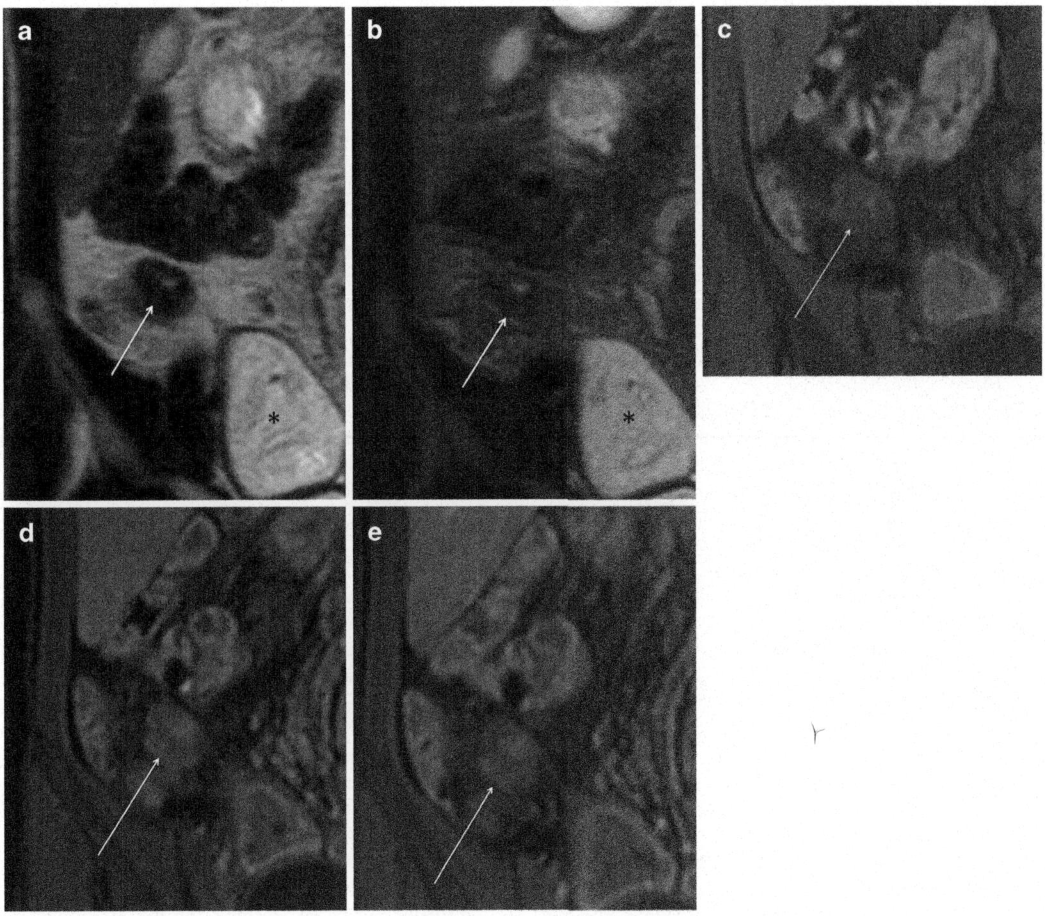

Fig. 5.32 Bowel wall thickening (arrow) is clearly seen on magnified GE T1-weighted fat-saturated image (**a**) and a slight hyperintensity is appreciable on fat-saturated T2-w image HASTE as well (**b**), which is due to coexistent mild inflammation of the bowel wall. Contrast-enhanced T1-weighted fat-saturated GE images obtained at arterial, portal, and delayed phases, respectively (**c**–**e**), show mild and initial mucosal enhancement (arterial phase) and delayed submucosal enhancement (delayed phase), a finding compatible with fibrotic degeneration

no correlation with histology for distinguishing inflammation from fibrosis [21, 54].

Finally, initial evidences on Magnetization Transfer (MT) emphasize the potential role of this sequence in distinguishing different degree of fibrosis even when inflammation coexists, outperforming ADC and contrast-enhanced scans [55].

References

1. Martin DR, Lauenstein T, Sitaraman SV. Utility of magnetic resonance imaging in small bowel Crohn's disease. Gastroenterology. 2007;133:385–90.
2. Satsangi J, Parkes M, Jewell DP, et al. Genetics of inflammatory bowel disease. Clin Sci (Lond). 1998;94:473–8.
3. Loftus EV Jr. Clinical epidemiology of inflammatory bowel disease: incidence, prevalence, and environmental influences. Gastroenterology. 2004;126:1504–17.
4. Bosani M, Ardizzone S, Porro GB. Biologic targeting in the treatment of inflammatory bowel diseases. Biologics. 2009;3:77–97.
5. Bansal P, Sonnenberg A. Risk factors for colorectal cancer in inflammatory bowel disease. Am J Gastroenterol. 1996;91:44–8.
6. Cheifetz AS. Management of active Crohn disease. JAMA. 2013;309:2150–8.
7. Del Vescovo R, Sansoni I, Caviglia R, et al. Dynamic contrast enhanced magnetic resonance imaging of the terminal ileum: differentiation of activity of Crohn's disease. Abdom Imaging. 2008;33:417–24.
8. Gourtsoyiannis NC, Grammatikakis J, Papamastorakis G, et al. Imaging of small intestinal Crohn's disease: comparison between MR enteroclysis and conventional enteroclysis. Eur Radiol. 2006;16:1915–25.
9. Masselli G, Brizi G, Parrella A, et al. Crohn disease magnetic resonance enteroclysis. Abdom Imaging. 2004;29:326–34.
10. Chou CK, Chen LT, Sheu RS, et al. MRI manifestations of gastrointestinal wall thickening. Abdom Imaging. 1994;19:389–94.
11. Prassopoulos P, Papanikolaou N, Grammatikakis J, et al. MR enteroclysis imaging of Crohn disease. Radiographics. 2001;21:s161–72.
12. Maselli G, Casciani E, Polettini, et al. Assessment of Crohn's disease in the small bowel: prospective comparison of magnetic resonance enteroclysis with conventional enteroclysis. Eur Radiol. 2006;16:2817–27.
13. Mazziotti S, Ascenti G, Scribano E, et al. Guide to magnetic resonance in Crohn's disease: from common findings to the more rare complicances. Inflamm Bowel Dis. 2011;17:1209–22.
14. Cicero G, Mazziotti S. Crohn's disease at radiological imaging: focus on techniques and intestinal tract. Intest Res. 2021;19(4):365–78. https://doi.org/10.5217/ir.2020.00097. Epub 2020 Nov 25. PMID: 33232590; PMCID: PMC8566824
15. Mazziotti S, Blandino A, Scribano E, Gaeta M, Mileto A, Fries W, Bombaci F, Ascenti G. MR enterography findings in abdominopelvic extraintestinal complications of Crohn's disease. J Magn Reson Imaging. 2013;37(5):1055–63. https://doi.org/10.1002/jmri.23859. Epub 2012 Oct 11. PMID: 23060240
16. Cicero G, Ascenti G, Bottari A, Catanzariti F, Blandino A, Mazziotti S. MR enterography: what is next after Crohn's disease? Jpn J Radiol. 2019;37(7):511–7. https://doi.org/10.1007/s11604-019-00838-y. Epub 2019 Apr 9. PMID: 30968265
17. Cicero G, Blandino A, D'Angelo T, Booz C, Vogl TJ, Ascenti G, Mazziotti S. Mimicking conditions of intestinal Crohn's disease: magnetic resonance enterography findings. Jpn J Radiol. 2022;40(1):19–28. https://doi.org/10.1007/s11604-021-01177-7. Epub 2021 Jul 25. PMID: 34304381
18. Tolan JM, Greenhalgh R, Zealley IA, et al. MR enterographic manifestation of small bowel Crohn disease. Radiographics. 2010;30:367–84.
19. Sempere GA, Martinez SV, Medina CE. MRI evaluation of inflammatory activity in Crohn's disease. Am J Roentgenol. 2005;184:1829–35.
20. Zappa M, Stefanescu C, Cazals-Hatem D, et al. Which magnetic resonance findings accurately evaluate inflammation in small bowel Crohn's disease? A retrospective comparison with surgical pathologic analysis. Inflamm Bowel Dis. 2011;17:984–93.
21. Wilkens R, Hagemann-Madsen RH, Peters DA, et al. Validity of contrast-enhanced ultrasonography and dynamic contrast-enhanced MR enterography in the assessment of transmural activity and fibrosis in Crohn's disease. J Crohns Colitis. 2018;12:48–56.
22. Tielbeek JA, Ziech ML, Li Z, et al. Evaluation of conventional, dynamic contrast enhanced and diffusion weighted MRI for quantitative Crohn's disease assessment with histopathology of surgical specimens. Eur Radiol. 2014;24:619–29.
23. Griffin N, Grant LA, Anderson S, et al. Small bowel MR enterography: problem solving in Crohn's disease. Insights Imaging. 2012;3:251–63.
24. Sinha R, Murphy P, Hawker P, et al. Role of MRI in Crohn's disease. Clin Radiol. 2009;64:341–52.
25. Oto A, Kulkarni K, Karczmar GS, et al. Evaluation of diffusion-weighted MR imaging for detection of bowel inflammation in patients with Crohn's disease. Acad Radiol. 2009;16:597–603.
26. Kiryu S, Dodanuki K, Takao H, et al. Free-breathing diffusion weighted imaging for the assessment of inflammatory activity in Crohn's disease. J Magn Reson Imaging. 2009;29:880–6.
27. Buisson A, Joubert A, Montoriol PF, et al. Diffusion-weighted magnetic resonance imaging for detecting and assessing ileal inflammation in Crohn's disease. Aliment Pharmacol Ther. 2013;37:537–45.

28. Froelich JM, Waldherr C, Stoupis C, et al. MR motility in Crohn's disease improves lesion detection compared with standard MR imaging. Eur Radiol. 2010;20:1945–51.
29. Odille F, Menys A, Ahmed A, et al. Quantitative assessment of small bowel motility by nonrigid registration of dynamic MR images. Magn Reson Med. 2012;68:783–93.
30. Menys A, Atkinson D, Odille F, et al. Quantified terminal ileal motility during MR enterography as a potential biomarker of Crohn's disease activity: a preliminary study. Eur Radiol. 2012;22:2494–501.
31. Yacoub HJ, Obara P, Oto A. Evolving role of MRI in Crohn's disease. J Magn Reson Imaging. 2013;37:1277–89.
32. Sinha R, Rajia P, Murphy P, et al. Utility of high-resolution MR imaging in demonstrating transmural pathologic changes in Crohn disease. Radiographics. 2009;29:1847–67.
33. Sinha R, Murphy P, Sanders S, et al. Diagnostic accuracy of high-resolution MR enterography in Crohn's disease: comparison with surgical and pathological specimen. Clin Radiol. 2013;68:917–27.
34. Guglielmo FF, Anupindi SA, Fletcher JG, Al-Hawary MM, Dillman JR, Grand DJ, Bruining DH, Chatterji M, Darge K, Fidler JL, Gandhi NS, Gee MS, Grajo JR, Huang C, Jaffe TA, Park SH, Rimola J, Soto JA, Taouli B, Taylor SA, Baker ME. Small bowel Crohn disease at CT and MR enterography: imaging atlas and glossary of terms. Radiographics. 2020;40(2):354–75. https://doi.org/10.1148/rg.2020190091. Epub 2020 Jan 17. PMID: 31951512
35. Laghi A, Paolantonio P, Catalano C, et al. MR imaging of the small bowel using polyethylene glycol solution as an oral contrast agent in adults and children with celiac disease: preliminary observations. Am J Roentgenol. 2003;180:191–4.
36. Low RN, Francis IR, Politoske D, et al. Crohn's disease evaluation: comparison of contrast-enhanced MR imaging and single-phase helical CT scanning. J Magn Reson Imaging. 2000;11:127–35.
37. Maccioni F, Viscido A, Broglia L, et al. Evaluation of Crohn disease activity with magnetic resonance imaging. Abdom Imaging. 2000;25:219–28.
38. Laghi A, Borrelli O, Paolantonio P, et al. Contrast enhanced magnetic resonance imaging of the terminal ileum in children with Crohn's disease. Gut. 2003;52:393–7.
39. Taylor SA, Punwani S, Rodriguez-Justo M, et al. Mural Crohn disease: correlation of dynamic contrast-enhanced MR imaging findings with angiogenesis and inflammation at histologic examination. Pilot Study Radiol. 2009;251:369–79.
40. Giusti S, Faggioni L, Neri E, et al. Dynamic MRI of the small bowel: usefulness of quantitative contrast-enhancement parameters and time-signal intensity curves for differentiating between active and inactive Crohn's disease. Abdom Imaging. 2010;35:646–53.
41. Oto A, Fan X, Mustafi D, et al. Quantitative analysis of dynamic contrast enhanced MRI for assessment of bowel inflammation in Crohn's disease: pilot study. Acad Radiol. 2009;16:1223–30.
42. Maccioni F, Bruni A, Viscido A, et al. MR imaging in patients with Crohn disease: MR sequences with the use of an oral superparamagnetic contrast agent. Radiology. 2006;238:157–530.
43. Gourtsoyianni S, Papanikolaou N, Amanakis E, et al. Crohn's disease lymphadenopathy: MR imaging findings. Eur J Radiol. 2009;69:425–8.
44. Gourtsoyiannis N, Papanikolaou N, Grammatikakis J, Papamastorakis G, Prassopoulos P, Roussomoustakaki M. Assessment of Crohn's disease activity in the small bowel with MR and conventional enteroclysis: preliminary results. Eur Radiol. 2004;14(6):1017–24. https://doi.org/10.1007/s00330-004-2302-8. Epub 2004 Apr 1
45. Church PC, Turner D, Feldman BM, et al. Systematic review with meta-analysis: magnetic resonance enterography signs for the detection of inflammation and intestinal damage in Crohn's disease. Aliment Pharmacol Ther. 2015;41:153–66.
46. Radmard AR, Eftekhar Vaghefi R, Montazeri SA, Naybandi Atashi S, Hashemi Taheri AP, Haghighi S, Salehnia A, Dadgostar M, Malekzadeh R. Mesenteric lymph nodes in MR enterography: are they reliable followers of bowel in active Crohn's disease? Eur Radiol. 2018;28(10):4429–37. https://doi.org/10.1007/s00330-018-5441-z. Epub 2018 Apr 25. PMID: 29696432
47. Gasche C, Scholmerich J, Brynskov J, et al. A simple classification of Crohn's disease: report of the working party for the world congress of gastroenterology, Vienna 1998. Inflamm Bowel Dis. 2000;6:8–15.
48. Satsangi J, Silverberg MS, Vermeire S, et al. The Montreal classification of inflammatory bowel disease: controversies, consensus, and implications. Gut. 2006;55:749–53.
49. Leyendecker JR, Bloomfeld RS, DiSantis DF, et al. MR enterography in the management of patients with Crohn disease. Radiographics. 2009;29:1827–46.
50. Schmidt S, Chevvalier P, Bessoud B, et al. Diagnostic performance of MRI for detection of intestinal fistulas in patients with complicated inflammatory conditions. Eur Radiol. 2007;17:2957–63.
51. Molendijk I, Peeters KC, Baeten CI, Veenendaal RA, van der Meulen-de Jong AE. Improving the outcome of fistulising Crohn's disease. Best Pract Res Clin Gastroenterol. 2014;28(3):505–18. https://doi.org/10.1016/j.bpg.2014.04.011. Epub 2014 May 5. PMID: 24913389
52. Rimola J, Planell N, Rodríguez S, et al. Characterization of inflammation and fibrosis in Crohn's disease lesions by magnetic resonance imaging. Am J Gastroenterol. 2015;110:432–40.
53. Rimola J, Capozzi N. Differentiation of fibrotic and inflammatory component of Crohn's disease-

associated strictures. Intest Res. 2020;18(2):144–50. https://doi.org/10.5217/ir.2020.00015. Epub 2020 Apr 20. PMID: 32326668; PMCID: PMC7206345

54. Knuesel PR, Kubik RA, Crook DW, Eigenmann F, Froehlich JM. Assessment of dynamic contrast enhancement of the small bowel in active Crohn's disease using 3D MR enterography. Eur J Radiol. 2010;73(3):607–13. https://doi.org/10.1016/j.ejrad.2008.12.001. Epub 2009 Jan 7. PMID: 19131201

55. Li XH, Mao R, Huang SY, Sun CH, Cao QH, Fang ZN, Zhang ZW, Huang L, Lin JJ, Chen YJ, Rimola J, Rieder F, Chen MH, Feng ST, Li ZP. Characterization of degree of intestinal fibrosis in patients with Crohn disease by using magnetization transfer MR imaging. Radiology. 2018;287(2):494–503. https://doi.org/10.1148/radiol.2017171221. Epub 2018 Jan 19. PMID: 29357272; PMCID: PMC6911697

6 Extraintestinal Complications in Crohn's Disease: MR-Enterography Findings

Tommaso D'Angelo, Silvio Mazziotti, Alfredo Blandino, and Giuseppe Cicero

Several extraintestinal complications have been associated to IBD. In particular, it has been noted that up to 36% of patients with CD have at least one extraintestinal manifestation and 25% have more than one extraintestinal complication. Moreover, the development of one extraintestinal complication appears to increase the risk of developing others [1–3].

This chapter discusses the usefulness of MRE even in detecting extraintestinal complications, which may sometimes be the initial presenting symptoms or, most commonly, be hidden by the major intestinal symptoms.

Abdominopelvic extraintestinal complications of CD, which can involve the hepatobiliary and pancreatic system, genitourinary system, musculoskeletal and cutaneous systems, as well as the peritoneum, can be included in an MR exam of the small bowel [4].

Different factors may be responsible for extraintestinal organ involvement, and this can make difficult the differentiation between true extraintestinal manifestations (i.e., primary systemic affections directly caused by IBD) and secondary extraintestinal complications of the disease, caused, for example, by malnutrition, chronic inflammation, or side effects of therapy.

Furthermore, even if some of these complications may not correlate with disease activity (e.g., primary sclerosing cholangitis and ankylosing spondylitis), generally, they tend to follow the clinical course of the disease and may have a high impact on the patient's life quality, morbidity, and also mortality [4, 5].

6.1 Hepatobiliary Complications

Hepatobiliary manifestations of IBD are not uncommon and include a variety of different conditions, such as primary sclerosing cholangitis, gallstone disease, liver abscess, and portal vein thrombosis [6, 7].

Even if rare, the involvement of the biliary tract can also be caused by the presence of inflammatory disease in the duodenum or by inflammation of the lesser omentum, which can lead to obstruction of the common bile duct, to ampullary stenosis (Figs. 6.1 and 6.2), or to duodenobiliary fistulas (Fig. 6.3).

6.1.1 Primary Sclerosing Cholangitis

One of the most serious complications of CD is primary sclerosing cholangitis (PSC), a disorder of both intrahepatic and extrahepatic bile ducts

T. D'Angelo · S. Mazziotti · A. Blandino
G. Cicero (✉)
Department of Biomedical Sciences and Morphological and Functional Imaging, University Hospital Messina, Messina, Italy
e-mail: tdangelo@unime.it; smazziotti@unime.it; ablandino@unime.it; gcicero@unime.it

S. Mazziotti et al. (eds.), *MR Enterography*, https://doi.org/10.1007/978-3-031-11930-9_6

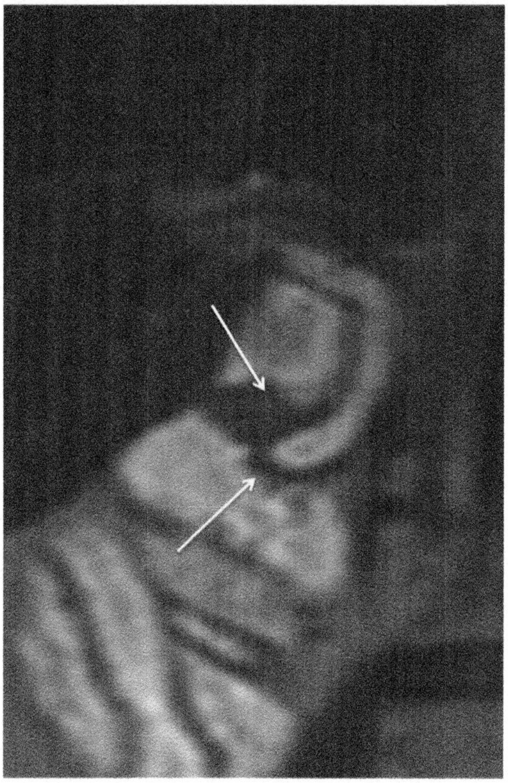

Fig. 6.1 Ampullary stenosis. Magnified coronal T2-w HASTE image shows wall-thickening in the ampullary area (arrows) with a mild upstream dilation of the common bile duct

[8]. It has an unclear etiology and a chronic and progressive course, manifesting itself as inflammation and fibrosis of the medium and large intra- and extrahepatic bile ducts.

Although most frequently found in patients suffering from ulcerative colitis (UC), PSC may occur in up to 4% of patients with CD, usually in those with colonic involvement and more likely in men than women [9–11].

PSC exposes the patient to the risk of cholangiocarcinoma development.

Moreover, an overlap between PSC and autoimmune hepatitis have been recorded in IBD patients, although mainly affected by UC [2, 10].

The MRE protocols, i.e., the MR fluoroscopy, well depict the typical findings of PSC, including multifocal strictures and irregularity of both intra- and extrahepatic bile ducts, leading to the classic "beads on a string" appearance (Fig. 6.4) [5, 12, 13]. It is the greatest risk factor for developing cholangiocarcinoma (they may present similar imaging characteristics), from which should be differentiated.

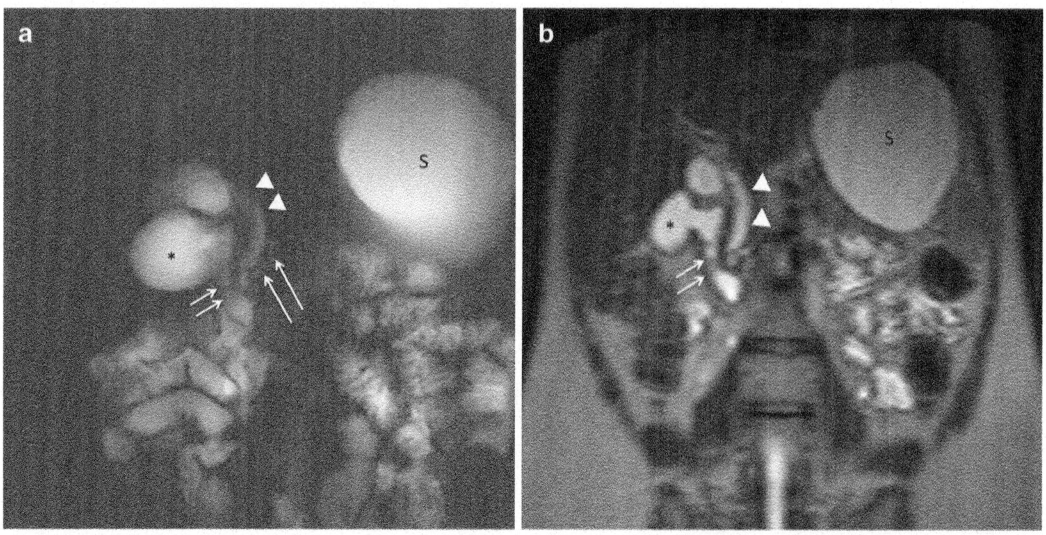

Fig. 6.2 Ampullary stenosis. Coronal T2-w RARE thick-slab (**a**) and coronal T2-w HASTE (**b**) images show the narrowing of duodenal lumen (short arrows in **a**), due to a concentric thickening of the duodenal wall in proximity of the ampullary area (short arrows in **b**). The consequent dilatation of the proximal duodenal bulb (asterisk) and of the stomach (S) can also be seen. The common bile duct (arrowheads) and pancreatic duct (long arrows in **a**) are visibly dilated

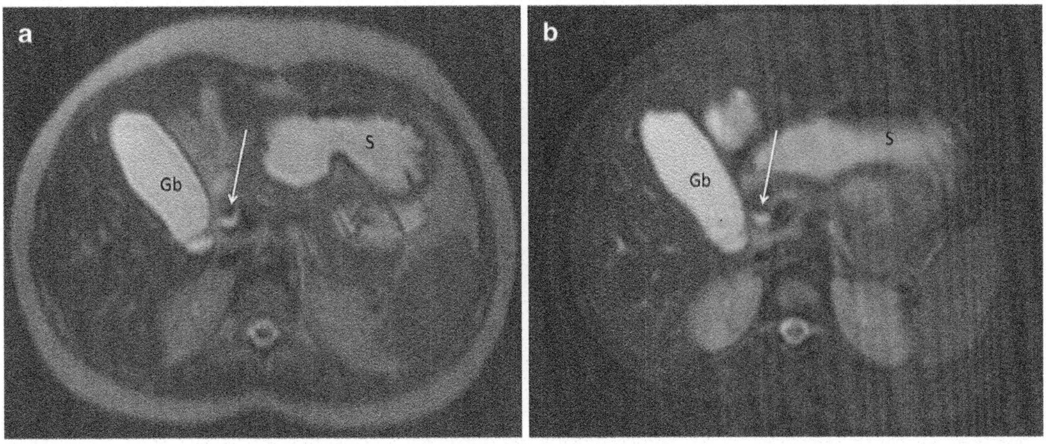

Fig. 6.3 Duodeno-biliary fistula. Axial T2-w HASTE (**a**) and axial T2-w fat-saturated HASTE (**b**) scans were obtained along the same anatomical plan. Both MR images show the presence of an air-fluid level in the common bile duct (arrow), due to an enterobiliary fistula. S: stomach; Gb: gallbladder

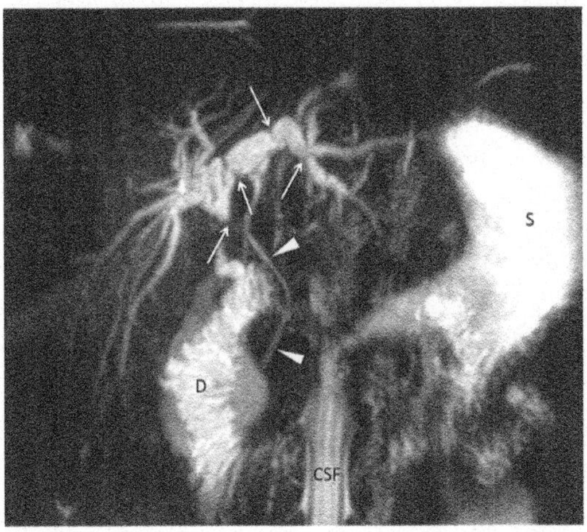

Fig. 6.4 Sclerosing cholangitis. Coronal MIP reformatted image, obtained from multislices T2-w fat-saturated HASTE sequence. Both the dilatations and the strictures of intra- and extrahepatic bile ducts (arrows) are well seen. Arrowheads: common bile duct; S: stomach; D: duodenum; CSF: cerebrospinal fluid

6.1.2 Gallstone Disease

Cholelithiasis is common in IBD patients, especially in those with CD [14].

The reported prevalence of gallstone disease (GD), defined as current gallstones or previous cholecystectomy for gallstones, in patients with CD, ranges from 13 to 34% [2, 13, 15].

In these patients, age, the site of CD at diagnosis, the number and sites of previous resections, as well as fasting and total parenteral nutrition are all independently associated with cholelithiasis, the pathogenesis of which is multifactorial.

The most important predisposing factor is malabsorption of bile acids from the inflamed terminal ileum, which leads to hepatic excretion of bile highly saturated with cholesterol; a further contributory role in the formation of gallstones could be played by a reduced and impaired fatty meal-induced gallbladder motility, evidenced in most of patients with ileal and ileocolonic disease [15, 16]. Another possibility may be a decreased release and/or hypersecretion of hormones stimulating (e.g., cholecystokinin) or inhibiting (e.g., somatostatin) gallbladder motility [17].

The CD duration is another important risk factor for GD, as well as previous surgery and number of resections are, with particular regard to those involving the ileocecal region.

The presence of cholelithiasis can be well detected at MR of the small bowel on both T2-w HASTE sequences and MR hydrographic images (Fig. 6.5).

6.1.3 Liver Abscess

Most liver abscesses usually occur in patients who have a long-standing history of inflamma-

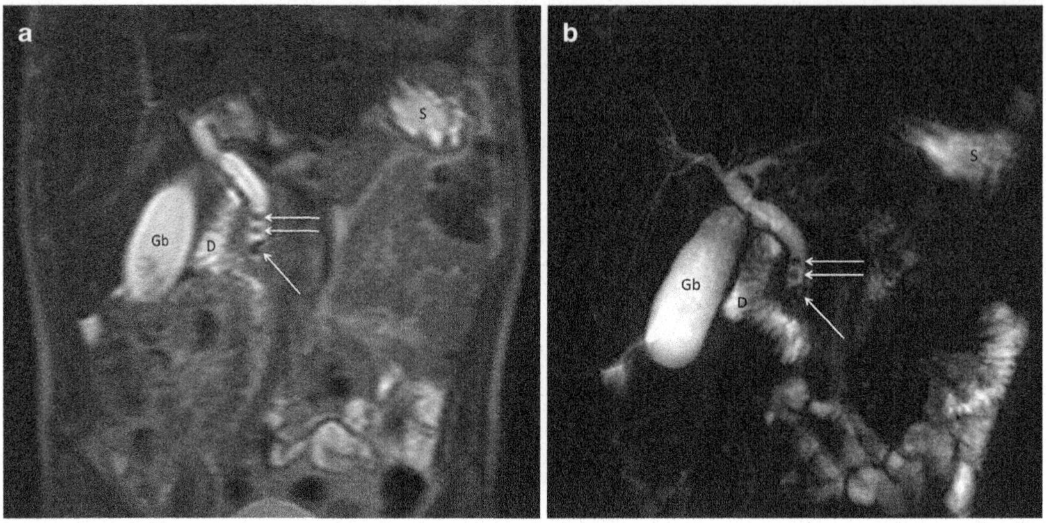

Fig. 6.5 Cholelithiasis. Coronal T2-w HASTE (**a**) and coronal T2-w RARE thick-slab (**b**) acquisitions show the presence of small filling defects within the common bile duct, consistent with gallstones (arrows). The cholesterol composition of the CBD gallstones was demonstrated after removal by means of retrograde endoscopic cholangiopancreatography. Gb: gallbladder; S: stomach. D: duodenum

tory bowel disease, although young patients can also be affected with multiple localizations [10, 18, 19]. Long-term steroid treatment, malnutrition, immunological abnormalities, and surgical interventions are additional predisposing factors that may lead to this occurrence.

The pathogenesis of liver abscesses has been linked to the interruption of normal mucosal integrity by the intestinal ulcerations with microbial invasion of the mesenteric vessels [2].

In fact, the invasion of portal venous system by microorganisms facilitates these pathogens to reach the liver.

Clinical manifestations of liver abscesses are variable, and common symptoms are fever, anorexia, weight loss, and abdominal pain that may often be misdiagnosed, as their clinical presentation can resemble an exacerbation of CD.

MR findings characteristic of pyogenic abscesses are high signal intensity on T2-weighted images and low signal intensity on T1-weighted images. Pyogenic abscesses usually possess markedly thick walls and internal septations, which enhance moderately to intensely on early-phase images and demonstrate persistent enhancement on late-phase images (Fig. 6.6).

6.1.4 Portal Vein Thrombosis

A rare complication in the nonsurgical clinical setting is represented by portal vein thrombosis.

Generally, patients with IBD are three times more likely to experience thromboembolic events, compared with the general population, with incidence reported to be between 1% and 41% [13].

The exact cause of the prothrombotic state is still unknown, even if it is likely to be multifactorial. Risk factors include active inflammation, immobility, surgery, and central venous catheter positioning as well as IBD itself.

During MRE, the signs of abdominal thromboembolic complication can be recognized as a filling defect within the portal lumen, especially on post-contrast images (Fig. 6.7) [2, 5].

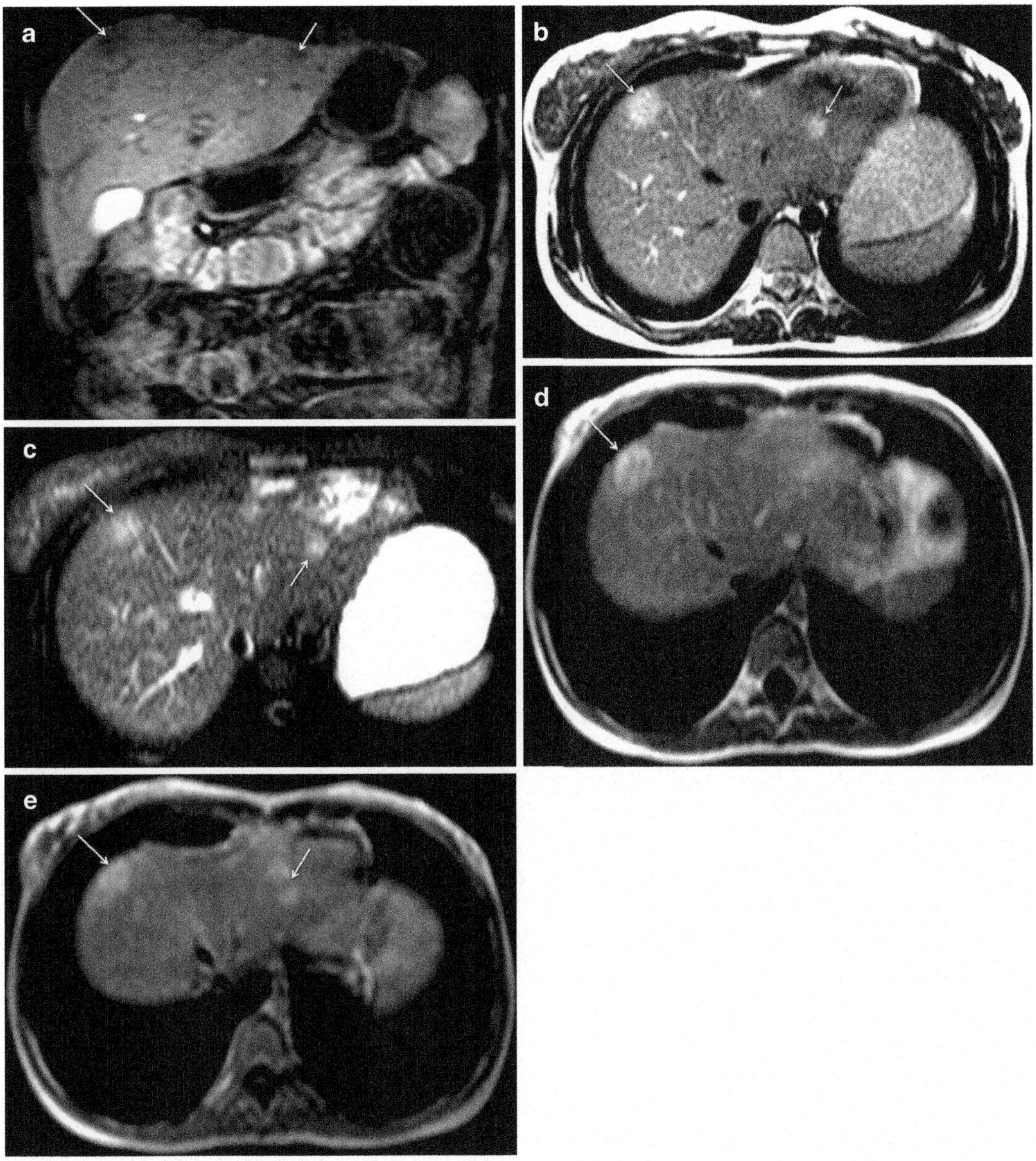

Fig. 6.6 Pyogenic liver abscesses in a febrile patient with Crohn's disease. Coronal T1-weighted fat-saturated GE image (**a**). Axial T2-w HASTE (**b**) and axial T2-w fat-saturated HASTE (**c**) scans. Axial T1-weighted fat-saturated GE images, obtained after i.v. administration of Gadolinium (**d**, **e**). MR shows two peripheral hepatic abscesses (arrows) hypointense on T1-weighted image (a), hyperintense on T2-weighted sequences (**b**, **c**) and characterized by marked peripheral enhancement on T1-weighted images obtained after contrast-medium administration (**d**, **e**). Follow-up MR examination, performed after antibiotic therapy, demonstrated the complete resolution

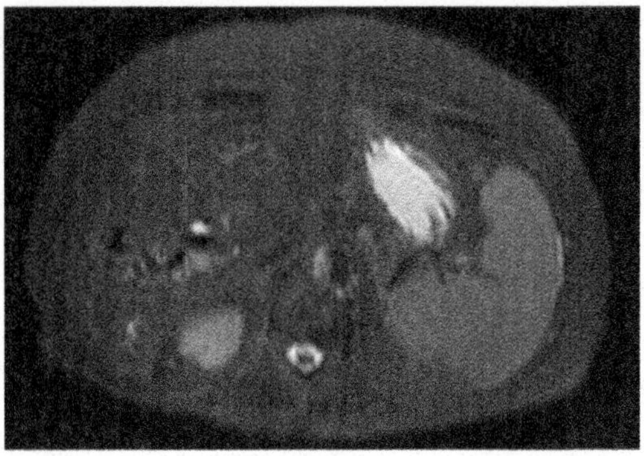

Fig. 6.7 Portal vein thrombosis. Axial T2-w fat-saturated HASTE image depicts right portal vein thrombosis with signs of periportal cavernomatosis. Because the patient had contraindications to contrast-medium administration (allergy and nephropathy), diagnosis of thrombosis of portal vein was confirmed by means of ultrasonography

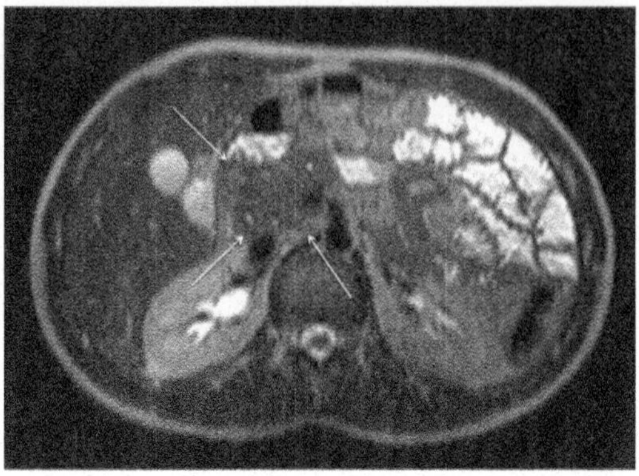

Fig. 6.8 CD patient in treatment with sulfasalazine and with a mild edematous form of head pancreatitis. Axial T2-w HASTE image shows mild enlargement of the pancreatic head (arrows) without any peripancreatic fluid collection

6.2 Pancreatic Complications

Acute and less often chronic pancreatitis may occur in patients affected by IBD as a result of the disease itself or secondary to medical treatment [20–22]. Many medications in IBD have the potential to induce acute pancreatitis; in particular, sulfasalazine, 5-aminosalicylic acid, azathioprine, 6-mercaptopurine, and rarely corticosteroids are well known to cause acute pancreatitis as a result of a possible idiosyncratic mechanism. Drug-induced pancreatitis typically occurs within the first weeks after drug therapy begins; the course is usually mild and resolves quickly after discontinuing the drug (Fig. 6.8).

Although rarely, autoimmune pancreatitis can occur in IBD subjects. The main correlation has been found between type-2 autoimmune pancreatitis and UC patients [23].

The IBD–pancreas association is further reflected in many reports of exocrine as well as endocrine pancreatic insufficiency.

The regional inflammatory complications due to duodenal and papillary involvement or biliary complications should also be considered when evaluating the pancreas at MRE.

For what concerns the pathogenesis of *chronic pancreatitis*, it still remains unclear. It has been considered to be caused by circulating inflammatory mediators rather than directly involved pancreatic tissue. Anyhow, autoantibodies against pancreatic tissue may also play a role in the development of exocrine insufficiency [24, 25].

MR findings include T2-hyperintensity on T2-weighted scans and loss of T1-hyperintensity on T1-weighted images, edema, or fluid collections within the surrounding fat tissue. Parenchymal or ductal calcifications can be detected as signal voids and are associated with chronic pancreatitis [2].

6.3 Genitourinary Complications

The topic of urogenital complications in inflammatory bowel disease is fairly narrow in its spectrum, and it is the result of anatomic and metabolic relationship between the two organ systems.

The most frequent conditions that fall under this list include ureteral obstruction, nephrolithiasis, fistulous disease involving the genitourinary tract, intrinsic renal disease associated with CD, and some considerations in those who have had surgical procedures that may have altered the normal pelvic anatomy.

6.3.1 Ureteral Obstruction

Obstructive uropathy can be found in up to 6% of CD patients [26].

Ureteral involvement is caused predominantly by transmural inflammation of the bowel, which can affect the retroperitoneum and results in ureter and bladder inflammation, fibrosis, or fistula formation [2, 4]. The right ureter is involved most commonly because of the high incidence of CD in the ileocecal region.

Urinary symptoms are often minimal because of the more evident intestinal symptoms, giving evidence of the important role of imaging for an early detection.

Heavily T2-weighted sequences, performed in MRE protocols, help to determine the degree and the location of ureteral obstruction, a valuable method to noninvasively assess the urinary tract of patients affected by CD. Nonetheless, T1-weighted sequences, performed after intravenous injection of contrast medium, can also provide information about the renal excretory function (Fig. 6.9).

6.3.2 Nephrolithiasis

Kidney stones are a common manifestation of CD, occurring in up to 12% of patients with IBD, and should always be a part of the differential diagnosis when the patient presents with localized pain [26].

Patients with severe intestinal involvement and a long disease course or who previously underwent extensive ileal resection or ileostomy formation are the ones with higher risk [27, 28]. The causes of lithogenesis in CD are believed to be dehydration associated with chronic diarrhea, decreased urine volume, aciduria, bowel resection, abnormal metabolism of oxalic acid, and oral administration of steroids or salazosulfapyridine [29].

In fact, in patients with IBD, there is a tendency for chronic volume contraction, due to loss of water and salt in diarrhea stool, which leads to decreased urine volumes. They also have a decreased intestinal absorption and a diminished urinary excretion of citrate and magnesium, which normally act as inhibitors of calcium oxalate crystallization.

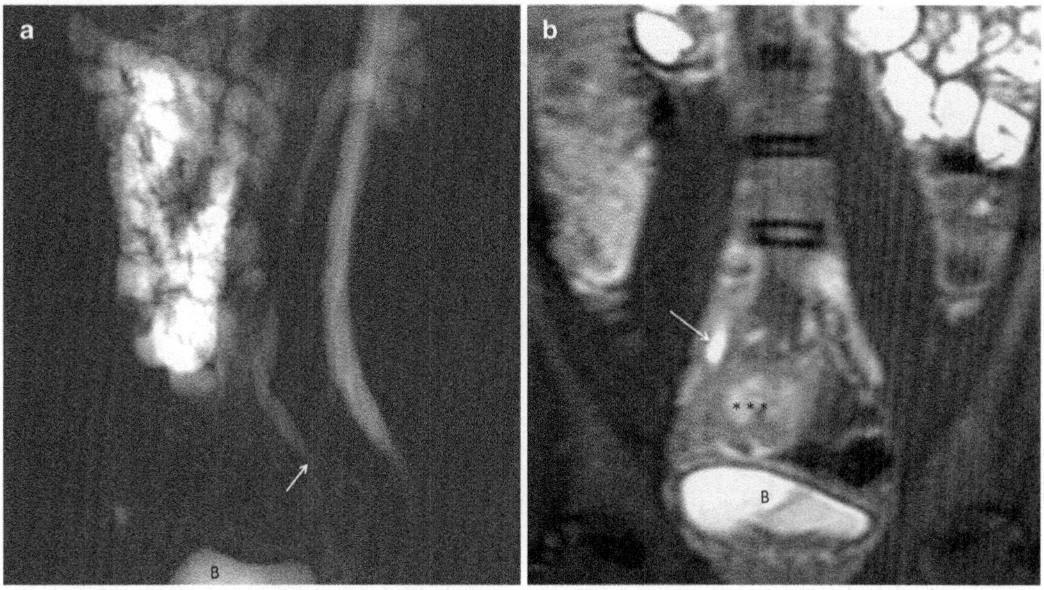

Fig. 6.9 Right ureterohydronephrosis due to pelvic ureter stenosis. Sagittal T2-w RARE thick-slab image (**a**) shows tapering of the pelvic ureter with mild degree of upstream dilation of ureteral and renal cavities (arrow). Coronal T2-w HASTE image (**b**) confirms the tapering of the ureter, which is caused by the diseased intestinal segment and the associated perivisceral inflammation (asterisks). B: bladder

Moreover, in the intestinal lumen of these patients, less oxalate is bound to luminal calcium, as the calcium preferentially binds to unabsorbed fatty acids, leading to an increased colonic absorption of oxalate and a decreased intraluminal component of insoluble calcium oxalate [2, 26].

For both these reasons, calcium oxalate stones are the most common type of stone encountered in patients with IBD, and for the same reasons, the treatment is supplemental calcium, rather than calcium restriction, as it is commonly and erroneously recommended.

Uric acid stones can also occur, usually in patients with ileostomies and high ileostomy outputs. In fact, in this case, there is a loss of a large volume of alkaline fluid from the stoma, which gives rise to a low volume of acidic urine favoring the precipitation of uric acid. To prevent this type of nephrolithiasis, the correct strategy is to increase the fluid intake, maintaining the patient on a low oxalate diet, and substituting medium-chain fatty acids for fats in the diet [27].

On MREg/MREc, ureteral dilation and luminal filling defect can be readily demonstrated with the use of heavily T2-weighted sequences, which are generally included in the standard protocol (Fig. 6.10).

6.3.3 Genitourinary Tract Fistulas

Patients with fistulous disease may present with urinary complaints prior to gastrointestinal abnormalities. It can involve several anatomical sites: upper urinary tract (kidney, ureter), lower urinary tract (bladder, urethra), or the female reproductive tract (vagina, uterus).

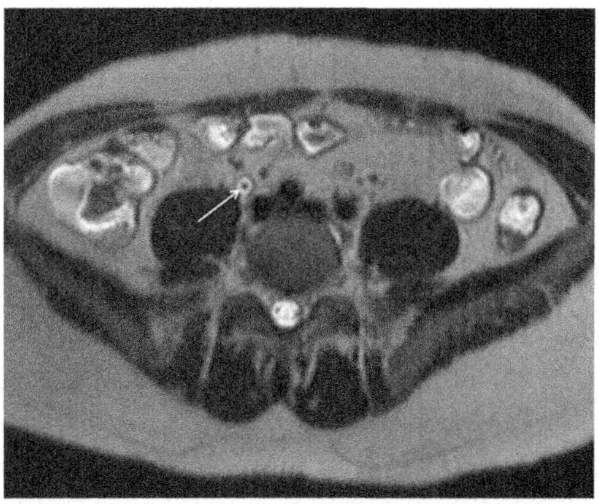

Fig. 6.10 Nephrolithiasis. Axial T2-w HASTE image shows a filling defect in the right ureter due to stone (arrow)

Enterovesical fistulas are the most common CD-related urinary fistula, affecting 2–8% of patients [30, 31].

They can arise as a manifestation of the transmural inflammation or as a complication of surgical resection, in the setting of anastomotic breakdown.

Even if it can be difficult to diagnose, the most common symptoms include pneumaturia, dysuria, and fecaluria. Enterovaginal fistulas can arise from the ileum or colon.

Usually, female patients are spared from enterovescical fistula formation due to the interposition of the uterus and adnexa [26].

Nevertheless, patients affected by CD who underwent hysterectomy appear to be associated with increased risk for the occurrence of a de novo fistula (Fig. 6.11).

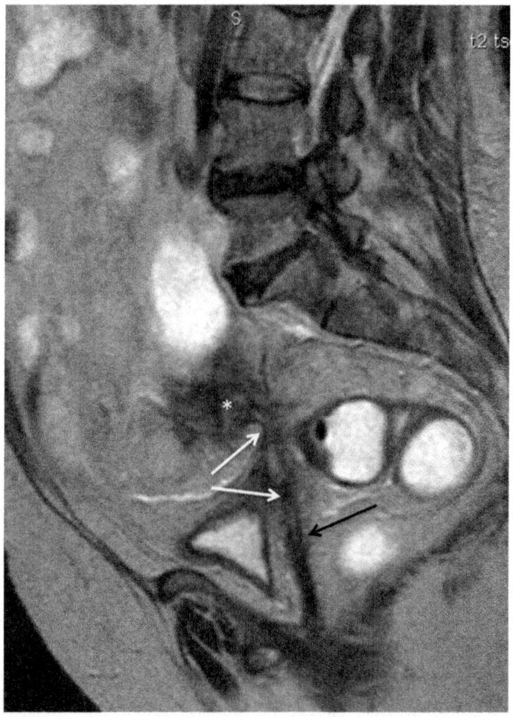

Fig. 6.11 Post-hysterectomy enterovaginal fistula. Sagittal T2-w HASTE image showing an enterovaginal fistula (white arrows). Black arrow: vaginal stump; asterisk: diseased ileal segment

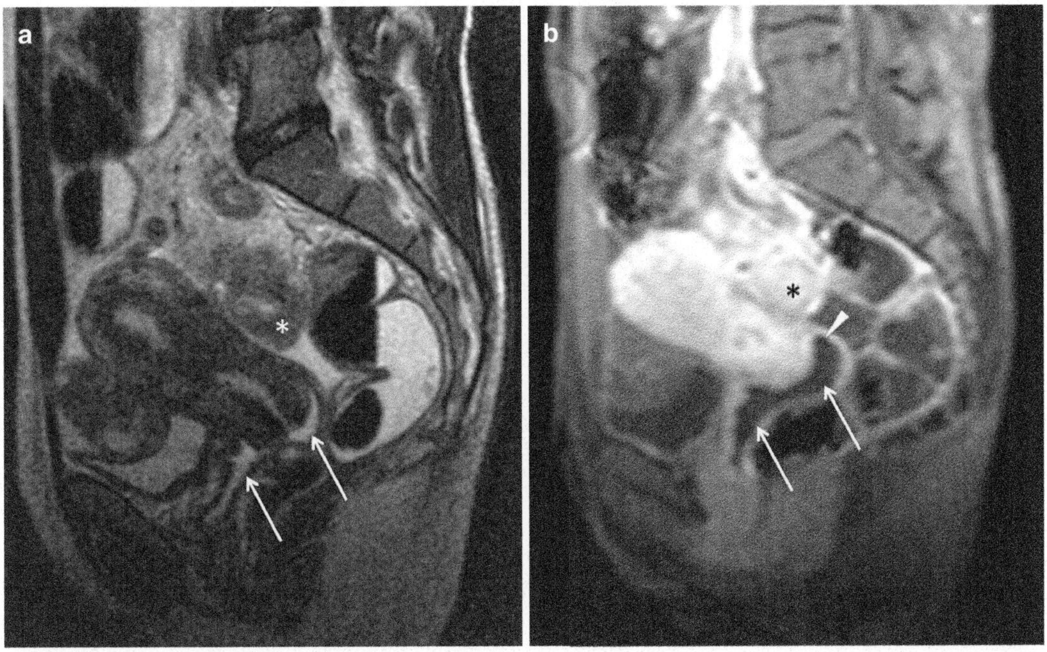

Fig. 6.12 Enterovaginal fistula. Sagittal T2-w HASTE image (**a**) shows the presence of fluid (arrows) within the vaginal lumen; on sagittal T1-w fat-saturated GE image obtained after i.v. administration of Gadolinium (**b**) and acquired 15 min after (**a**), a further filling of the vaginal cavity (arrows) is seen, as well as a gas bubble in the posterior vaginal fornix (arrowhead in **b**); both these signs are highly suggestive for enterovaginal fistula. Asterisk: inflamed bowel segment

Radiologists should be familiar with findings of genitourinary fistulas for both accurate diagnosis and, in many cases, guidance of management planning.

The direct identification of fistulas is not always possible, even if progressive filling with PEG-water solution of the uterine or vaginal cavity, observed at MRE performed with several sequential acquisitions, represents a reliable sign of enterovaginal or enterouterine fistula (Fig. 6.12).

6.4 Musculoskeletal and Cutaneous Manifestations

IBD-related arthropathy is part of a subset of diseases broadly termed "seronegative spondyloarthropathies." Arthritis occurs equally in males and females and is generally more common in patients with colonic disease than those with small bowel disease [32–34]. Approximately 12.8–23% of patients have peripheral arthritis that typically runs a parallel disease activity course to that of the bowel [35, 36]. Axial arthropathies instead are not associated with disease activity.

The articular involvement can precede, be synchronous with, or develop following the diagnosis of IBD [32–34].

Progressive ankylosing spondylitis and sacroiliitis, sometimes paucisymptomatic, may both occur.

MR is generally characterized by high sensitivity in detecting signs of sacroiliitis, and considering that as many as 27% of patients with CD eventually have radiographic evidence of sacroiliitis, at MRE, attention has to be paid to detect any early signs.

MR signs of sacroiliitis are characterized by increased signal intensity in the sacroiliac joints and subchondral bone marrow on fat-suppressed T2-weighted images, as well as by

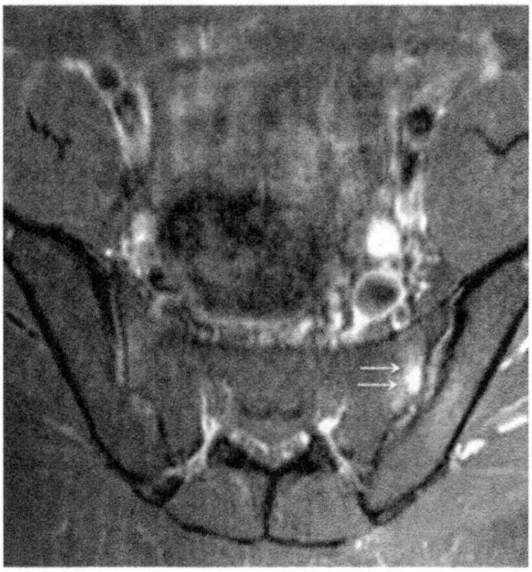

Fig. 6.13 Sacroiliitis. Axial T1-w fat-saturated GE image obtained after i.v. injection of Gadolinium shows enhancement of both the sacroiliac joint spaces, as well as a mild subchondral enhancement of sacral bone. Such findings are more evident on the left side (arrows)

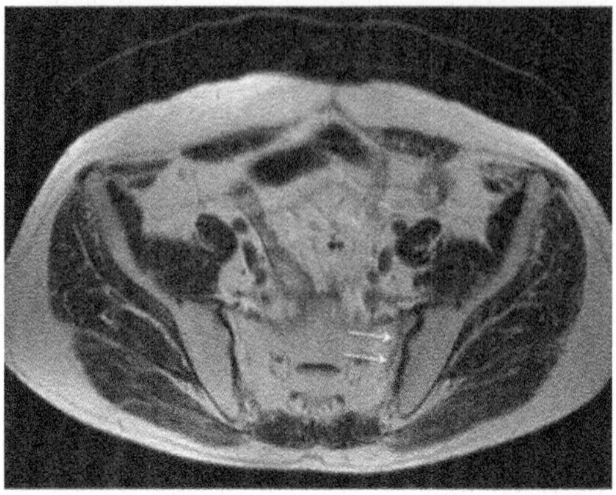

Fig. 6.14 Insufficiency fracture. Axial T2-w TSE image showing a linear hypointensity (arrows) in the left sacralis ala that runs parallel to sacroiliac joint

enhancement of the joint space and subchondral region on fat-suppressed T1-weighted images obtained after intravenous injection of contrast medium (Fig. 6.13).

Although it is not strictly targeted to the evaluation of the sacroiliac joints, MRE has been demonstrated to be an accurate tool for detection of early signs of sacroiliitis in pediatric patients, even when asymptomatic [37].

Osteoporosis is nowadays recognized as a common complication of IBD and in particular of CD. Beyond age-related risk factors that are present in the general population, there are multiple IBD-specific risk factors, including corticosteroid therapy, inflammatory-mediated bone resorption, calcium and magnesium dietary malabsorption, poor dietary calcium intake (related to lactose intolerance), vitamin D deficiency, decreased serum albumin levels, and reduced physical activity [2].

In particular, male patients with low body mass index and history of prior bowel resection are predominantly vulnerable to develop osteoporosis [38].

Patients with IBD are therefore at increased risk of sustaining insufficiency fractures at several sites, such as the spine and pelvis (Fig. 6.14).

Cutaneous manifestations have been reported in up to 43% of CD patients.

They may be contiguous to the gastrointestinal tract or distant.

The different cutaneous manifestations in IBDs may be grouped into four main entities: "disease specific," which share the same histopathological findings of IBDs (i.e., fistulas or metastatic Crohn's disease); "reactive," which are triggered by IBDs but without pathologic features in common (erythema nodosum and pyoderma gangrenosum); "associated," which are simultaneous but not caused by IBD (psoriasis, vitiligo, eczema); "drug-related" (eczema, psoriasis, skin infections, or cancers) [39–41].

Although these entities are usually recognized at physical examination, some of them, such as enterocutaneous fistulas, can be detected at MRE (Fig. 6.15).

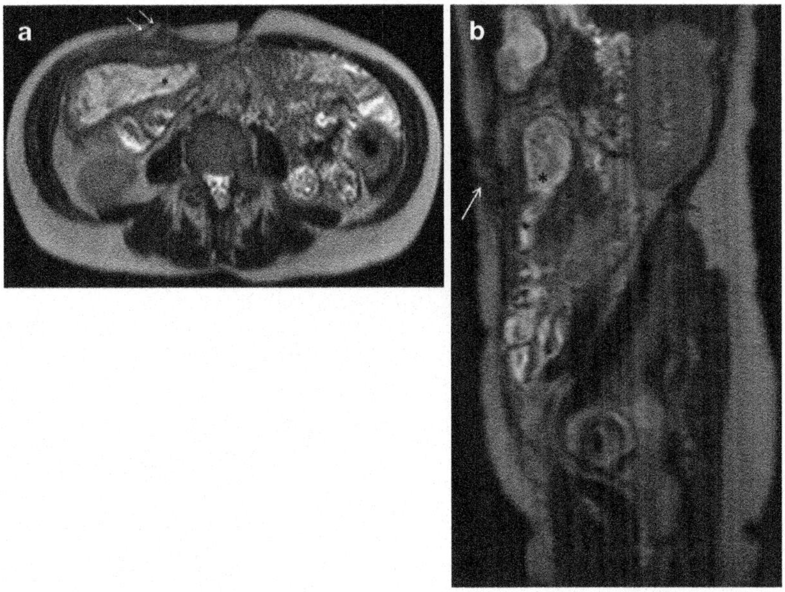

Fig. 6.15 Enterocutaneous fistula. Axial (**a**) and sagittal (**b**) T2-w HASTE sequences show an enterocutaneous fistula (arrows) extending from a pathologic ileal loop (black asterisk) to the cutaneous surface

6.5 Peritoneal Involvement

Normally, the peritoneum easily absorbs fluid; when inflamed or mechanically injured, the serosal properties are changed and fluid absorption is impaired. Moreover, in females, peritoneal fluid is formed by exudation from an active ovary as suggested by the higher concentration of ovarian steroid hormones in peritoneal fluid than in plasma [42, 43]. When chronic peritoneal inflammation or previous surgery leads to formation of adhesions that envelop an active ovary, ovarian fluids accumulate and a peritoneal pseudocyst may develop [44, 45].

MRE is particularly useful in detecting peritoneal pseudocysts in CD, especially on T2-weighted scans which are sensitive to fluids.

On MR imaging, pseudocysts appear irregular in shape, and the cyst walls may reflect the invaginations of the surrounding structures. MR characteristics of peritoneal inclusion cysts are low signal intensity on T1-weighted images and high signal intensity on T2-weighted spin-echo images, suggesting that the fluid is serous. T2-weighted sequences performed during MRE allow the correct diagnosis in the right clinical setting, useful for a prompt treatment considering the risk of onset after surgical resection (30–50%) (Fig. 6.16) [43, 46].

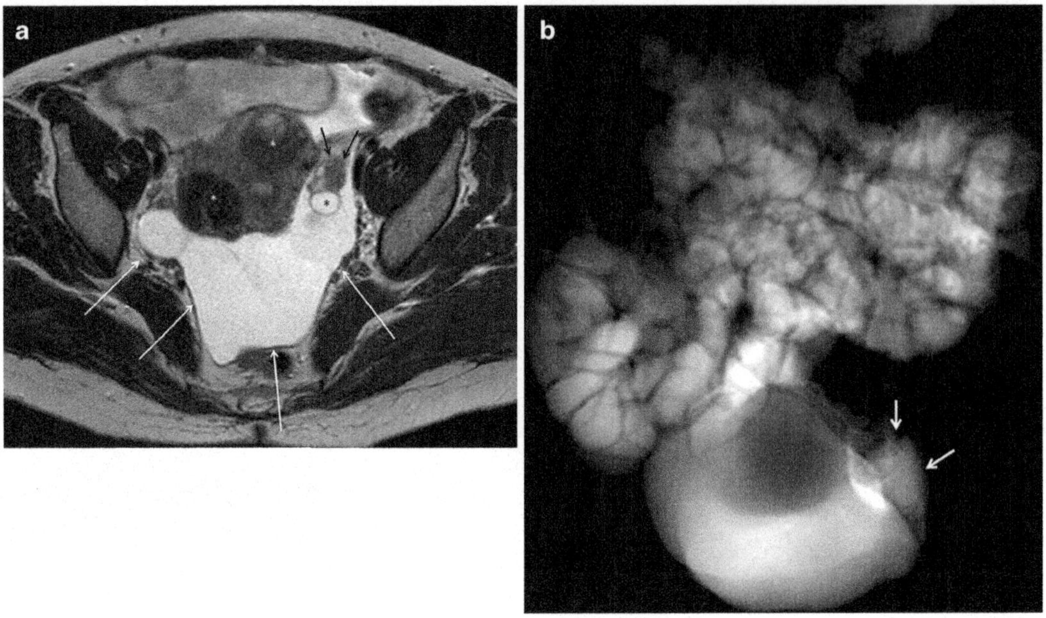

Fig. 6.16 Peritoneal pseudocyst. Axial T2-w TSE image (**a**) showing a large peritoneal fluid collection outlined by peritoneum (white arrows). Note the left ovary at the periphery of the peritoneal pseudocyst (black arrows) with a small functional cyst (black asterisk). Uterine fibroids (white asterisks) can be seen. The same findings (the large peritoneal pseudocyst and the ovary with the cystic lesion) are also well shown on coronal T2-w RARE thick-slab (**b**)

References

1. Bernstein CN, Blanchard JF, Rawsthorne P, et al. The prevalence of extra-intestinal disease in inflammatory bowel disease: a population based study. Am J Gastroenterol. 2001;96:1116–22.
2. Olpin JD, Sjoberg BP, Stilwill SE, Jensen LE, Rezvani M, Shaaban AM. Beyond the bowel: extraintestinal manifestations of inflammatory bowel disease. Radiographics. 2017;37(4):1135–60. https://doi.org/10.1148/rg.2017160121. Epub 2017 May 26. PMID: 28548906
3. Cicero G, Mazziotti S. Crohn's disease at radiological imaging: focus on techniques and intestinal tract. Intest Res. 2021;19(4):365–78. https://doi.org/10.5217/ir.2020.00097. Epub 2020 Nov 25. PMID: 33232590; PMCID: PMC8566824
4. Mazziotti S, Ascenti G, Scribano E, et al. Guide to magneticresonance in Crohn's disease: from common findings to the more rare complicances. Inflamm Bowel Dis. 2011;17:1209–22.
5. Mazziotti S, Blandino A, Scribano E, et al. MR enterography findings in abdominopelvic extraintestinal complications of Crohn's disease. J Magn Reson Imaging. 2013;37:1055–63.
6. Fousekis FS, Katsanos KH, Theopistos VI, Baltayiannis G, Kosmidou M, Glantzounis G, Christou L, Tsianos EV, Christodoulou DK. Hepatobiliary and pancreatic manifestations in inflammatory bowel diseases: a referral center study. BMC Gastroenterol. 2019;19(1):48. https://doi.org/10.1186/s12876-019-0967-3. PMID: 30943899; PMCID: PMC6446300
7. Ravindran S, Hancox SH, Barlow N, Dunk A, Howlett D. Unexpected findings in magnetic resonance enterography and their clinical significance. Can J Gastroenterol Hepatol. 2016;2016:4020569. https://doi.org/10.1155/2016/4020569. Epub 2016 Mar 29. PMID: 27446837; PMCID: PMC4904694
8. Wee A, Ludwig J, Hellers G, et al. Extracolonic diagnosis of ulcerative colitis: an epidemiological study. Am J Gastroenterol. 1990;85:711–6.
9. de Vries AB, Janse M, Blokzijl H, Weersma RK. Distinctive inflammatory bowel disease phenotype in primary sclerosing cholangitis. World J Gastroenterol. 2015;21(6):1956–71. https://doi.org/10.3748/wjg.v21.i6.1956. PMID: 25684965; PMCID: PMC4323476
10. Fousekis FS, Theopistos VI, Katsanos KH, Tsianos EV, Christodoulou DK. Hepatobiliary manifestations and complications in inflammatory bowel disease: a review. Gastroenterology Res. 2018;11(2):83–94. https://doi.org/10.14740/gr990w. Epub 2018 Apr 7. PMID: 29707074; PMCID: PMC59166314
11. Ramussen HH, Fallinborg JF, Mortensen PB, et al. Hepatobiliary dysfunction and primary sclerosing cholangitis in patients with Crohn's disease. Scand J Gastroenterol. 1997;32:604–10.

12. Alexopoulou E, Xenophontos PE, Economopoulos N, et al. Investigative MRI cholangiopancreatography for primary sclerosing cholangitis-type lesions in children with IBD. J Pediatr Gastroenterol Nutr. 2012;55:308–13.
13. Navaneethan U, Shen B. Hepatopancreatobiliary manifestations and complications associated with inflammatory bowel disease. Inflamm Bowel Dis. 2010;16:1598–619.
14. Zhang FM, Xu CF, Shan GD, Chen HT, Xu GQ. Is gallstone disease associated with inflammatory bowel diseases? A meta-analysis. J Dig Dis. 2015;16(11):634–41. https://doi.org/10.1111/1751-2980.12286. PMID: 26332254
15. Fraquelli M, Losco A, Visentin S, et al. Gallstone disease and related risk factors in patients with Crohn disease: analysis of 330 consecutive cases. Arch Intern Med. 2001;161:2201–4.
16. Dowling RH, Bell GD, White J. Lithogenic bile in patients with ileal dysfunction. Gut. 1972;13:415–20.
17. Fraquelli M, Bardella MT, Peracchi M, et al. Gallbladder emptying and somatostatin and cholecystokinin plasma levels in celiac disease. Am J Gastroenterol. 1999;94:1866–70.
18. Filik L, Ulker A, Tunc B, et al. Liver abscess: a rare but an important complication that must be considered in Crohn's disease. Acta Gastroenterol Belg. 2004;67:303–5.
19. Karaca C, Pinarbasi B, Danalioglu A, et al. Liver abscess as a rare complication of Crohn's disease: a case report. Turk J Gastroenterol. 2004;15:45–8.
20. Pitchumoni CS, Rubin A, Das K. Pancreatitis in inflammatory bowel diseases. J Clin Gastroenterol. 2010;44:246–53.
21. Moolsintong P, Loftus EV Jr, Chari ST, et al. Acute pancreatitis in patients with Crohn's disease: clinical features and outcomes. Inflamm Bowel Dis. 2005;11:1080–4.
22. Inoue H, Shiraki K, Okano H, et al. Acute pancreatitis in patients with ulcerative colitis. Dig Dis Sci. 2005;50:1064–7.
23. Ramos LR, Sachar DB, DiMaio CJ, Colombel JF, Torres J. Inflammatory bowel disease and pancreatitis: a review. J Crohns Colitis. 2016;10(1):95–104. https://doi.org/10.1093/ecco-jcc/jjv153. Epub 2015 Sep 7. PMID: 26351384
24. Ransford RA, Langman MJ. Sulphasalazine and mesalazine: serious adverse reactions re-evaluated on the basis of suspected adverse reaction reports to the Committee on Safety of Medicines. Gut. 2002;51:536–9.
25. Maconi G, Dominici R, Molteni M, et al. Prevalence of pancreatic insufficiency in inflammatory bowel diseases. Assessment by fecal elastase-1. Dig Dis Sci. 2008;53:262–70.
26. Tonolini M, Villa C, Campari A, Ravelli A, Bianco R, Cornalba G. Common and unusual urogenital Crohn's disease complications: spectrum of cross-sectional imaging findings. Abdom Imaging. 2013;38(1):32–41. https://doi.org/10.1007/s00261-012-9876-4. PMID: 22456714
27. Worcester EM. Stones from bowel disease. Endocrinol Metab Clin N Am. 2002;31:979–99.
28. McConnell N, Campbell S, Gillanders I, et al. Risk factors for developing renal stones in inflammatory bowel disease. BJU Int. 2002;89:835–41.
29. Stote RM, Smith LH, Dubb JW, et al. Oxypurinol nephrolithiasis in regional enteritis secondary to allopurinol therapy. Ann Intern Med. 1980;92:384–5.
30. Ishii G, Nakajima K, Tanaka N. Clinical evaluation of urolithiasis in Crohn's disease. Int J Urol. 2009;16:477–80.
31. Schwartz DA, Loftus EV, Tremaine WJ, et al. The natural history of fistulizing Crohn's disease in Olmsted County, Minnesota. Gastroenterology. 2002;122:875–80.
32. Herrmann KA, Michaely HJ, Zech CJ, et al. Internal fistulas in Crohn's disease: magnetic resonance enteroclysis. Abdom Imaging. 2006;31:675–87.
33. Peeters H, Vander Cruyssen B, Mielants H, et al. Clinical and genetic factors associated with sacroiliitis in Crohn's disease. J Gastroenterol Hepatol. 2008;23:132–7.
34. de Vlam K, Mielants H, Cuvelier C, et al. Spondyloarthropathy is underestimated in inflammatory bowel disease: prevalence and HLA association. J Rheumatol. 2000;27:2860–5.
35. Mielants H, Veys EM, De Vos M, et al. The evolution of spondyloarthropathies in relation to gut histology. Clinical aspects. J Rheumatol. 1995;22:2266–72.
36. Dekker-Saeys BJ, Meuwissen SGM, Van Den Berg-Loonen EM, et al. Prevalence of peripheral arthritis, sacroiliitis, and ankylosing spondylitis in patients suffering from inflammatory bowel disease. Ann Rheum Dis. 1978;37:36–41.
37. Giani T, Bernardini A, Basile M, Di Maurizo M, Perrone A, Renzo S, Filistrucchi V, Cimaz R, Lionetti P. Usefulness of magnetic resonance enterography in detecting signs of sacroiliitis in young patients with inflammatory bowel disease. Pediatr Rheumatol Online J. 2020;18(1):42. https://doi.org/10.1186/s12969-020-00433-w. PMID: 32493352; PMCID: PMC7268528
38. Hoffmann P, Krisam J, Kasperk C, Gauss A. Prevalence, risk factors and course of osteoporosis in patients with Crohn's disease at a tertiary referral Center. J Clin Med. 2019;8(12):2178. https://doi.org/10.3390/jcm8122178. PMID: 31835600; PMCID: PMC6947604
39. Ungureanu L, Cosgarea R, Alexandru Badea M, Florentina Vasilovici A, Cosgarea I, Corina ŞS. Cutaneous manifestations in inflammatory bowel disease (review). Exp Ther Med. 2020;20(1):31–7. https://doi.org/10.3892/etm.2019.8321. Epub 2019 Dec 12. PMID: 32508989; PMCID: PMC7271697
40. Gravina AG, Federico A, Ruocco E, Lo Schiavo A, Romano F, Miranda A, Sgambato D, Dallio M, Ruocco

V, Loguercio C, Romano M. Crohn's disease and skin. United European Gastroenterol J. 2016;4(2):165–71. https://doi.org/10.1177/2050640615597835. Epub 2015 Aug 21. PMID: 27087942; PMCID: PMC4804366
41. Hagen JW, Swoger JM, Grandinetti LM. Cutaneous manifestations of Crohn disease. Dermatol Clin. 2015;33(3):417–31. https://doi.org/10.1016/j.det.2015.03.007. PMID: 26143422
42. Tavarela Veloso F. Review article: skin complications associated with inflammatory bowel disease. Aliment Pharmacol Ther. 2004;20:50–3.
43. Maathuis JB, Van Look PFA, Michie EA. Changes in volume total protein and ovarian concentrations of peritoneal fluid throughout the human menstrual cycle. J Endocrinol. 1978;76:123–33.
44. Komickx PR, Renaer M, Brosens IA. Origin of peritoneal fluid in women: an ovarian exudation product. Br J Obstet Gynaecol. 1980;87:177–83.
45. Kim JS, Lee HJ, Woo SK, et al. Peritoneal inclusion cysts and their relationship to the ovaries: evaluation with sonography. Radiology. 1997;204:481–4.
46. Mazziotti S, D'Angelo T, Racchiusa S, Salamone I, Blandino A, Ascenti G. Peritoneal inclusion cysts in patients affected by Crohn's disease: magnetic resonance enterography findings in a case series. Clin Imaging. 2016;40(1):152–5. https://doi.org/10.1016/j.clinimag.2015.09.010. Epub 2015 Sep 18. PMID: 26456117

7 MR-Enterography beyond Crohn's Disease

Giuseppe Cicero, Tommaso D'Angelo, Giorgio Ascenti, and Silvio Mazziotti

The main clinical indication for MRE performance is represented by Crohn's disease (CD) [1].

Nevertheless, its increased worldwide use and the undoubted advantages consisting of high contrast resolution, lack of ionizing radiation, and comprehensive evaluation of the whole abdominal cavity, with possibility of bowel assessment and extraintestinal complications detection, have made this technique fascinating for the evaluation of different intestinal pathologies.

Therefore, the literature has been enriched by several MRE applications further than CD.

However, it is important to keep in mind that MRE findings specific to one intestinal disease have not been found yet.

On one hand, this means that MRE alone is not sufficient for a definite diagnosis, but it has to be complemented by clinical symptoms, laboratory tests, and endoscopy with biopsy.

On the other hand, it is mandatory for the radiologist to acknowledge the wide differential diagnosis of CD on MRE, since useful information can be provided not only for completing the patient clinical assessment but also for guiding the physicians to the right diagnosis.

7.1 Congenital Anomalies, Anatomical Variants, and Paraphysiological Findings

Malrotation of the small bowel is a type of congenital rotation and fixation anomaly that can lead to an acute midgut volvulus in newborns.

On the other hand, in adults it is usually detected as an incidental finding, although a clinical history of sub-occlusion episodes, with pain and vomit can be present.

MRI is not the first-line imaging modality in case of emergency.

However, when symptoms are subtle and chronic, MRI can take advantage of the enterographic protocol exploiting dilation of intestinal loops for small bowel assessment.

Main findings of malrotation include right-sided localization of the duodenojejunal flexure, dilation of the proximal duodenum, prevalent positioning of the small bowel loops within the right abdominal cavity, and inversed position of superior mesenteric artery and vein [2, 3].

Meckel diverticulum results from an incomplete obliteration of the omphalomesenteric duct and it can be detected as a blind tubular structure in continuity with the small bowel [3].

G. Cicero · T. D'Angelo · G. Ascenti
S. Mazziotti (✉)
Department of Biomedical Sciences and Morphological and Functional Imaging, University Hospital Messina, Messina, Italy
e-mail: gcicero@unime.it; tdangelo@unime.it; gascenti@unime.it; smazziotti@unime.it

S. Mazziotti et al. (eds.), *MR Enterography*, https://doi.org/10.1007/978-3-031-11930-9_7

Meckel diverticulitis can manifest with symptoms similar to appendicitis and can be further complicated by bleeding and perforation [4, 5].

Especially when inflamed, at MRE Meckel diverticulum can show wall thickening, luminal dilation, and reactive inflammation of the adjacent small bowel loops [3, 5].

Foregut duplication cysts are congenital abnormalities that can arise in any part of the gastrointestinal tract, with a higher incidence at the level of ileum or ileocecal valve [3, 5].

The features are those of a cystic lesion with a peripheral thin wall and a variable content, for instance, due to hemorrhage or infections [3, 5].

When large in size, a palpable mass can be appreciated and an obstructive effect can be determined on the adjacent loops due to extrinsic compression or intussusception [3, 5].

Terminal ileum can show an abnormal thickening in lymphoid nodular hyperplasia, a normal variant in the pediatric population and young adults that cannot be distinguished from Crohn's disease on T2-weighted, contrast-enhanced, and DWI sequences [6].

In adults, lipomatosis can cause a prominent and thickened aspect of the ileocecal valve due to adipose infiltration. In this case T2-weighted sequences with fat saturation can be helpful in showing drop of intensity signal (Fig. 7.1). Supplementary acquisition of T1-weighted sequences without fat suppression, generally not included in the standard MRE protocol, can be helpful in demonstrating the fat content [7].

7.2 Acquired Nonneoplastic Intestinal Conditions

7.2.1 Motility Disorders

Bowel obstruction is an emergency condition that may require surgery and it is usually assessed through abdominal x-ray plain films and CT scan.

MRI can be performed as an alternative technique in order to spare radiation exposure.

A number of pathologies can cause obstruction, classified as extrinsic (i.e., volvulus, hernias, and extraintestinal abdominal masses), intrinsic (i.e., malignancies and radiation enteritis), or intraluminal (i.e., foreign bodies and bezoar) [8].

MRE exploits the intestinal distention through the assumption of oral contrast medium in order to identify the transition point, which results in an abrupt narrowing of loop caliber with dilation of the upstream intestinal segments [5, 8, 9].

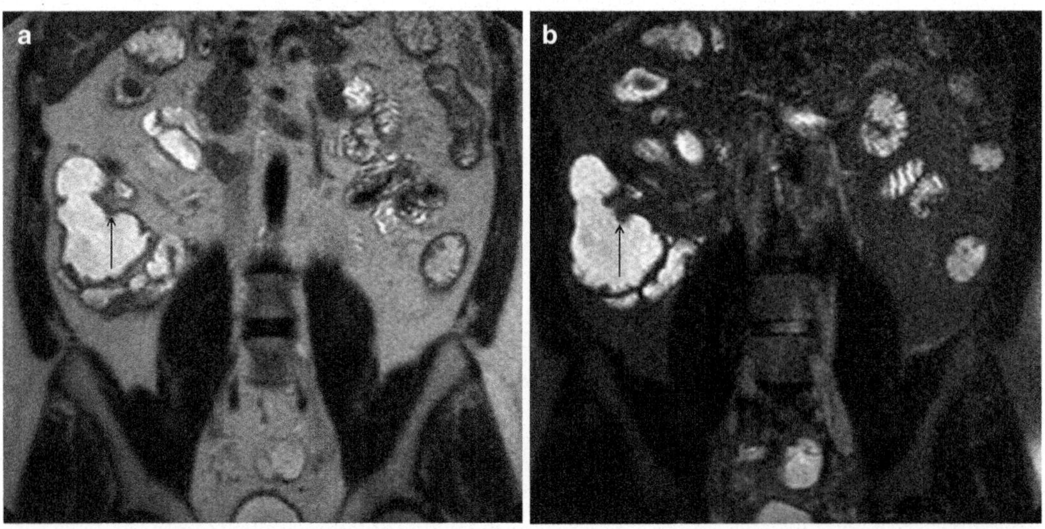

Fig. 7.1 A thickening of the ileocecal valve (arrows) is identifiable at coronal T2-w HASTE scan (**a**). The thickening shows a drop of signal on coronal T2-w fat-saturated HASTE acquisition (**b**) referrable to lipomatosis

In patients who underwent prior surgery, detection of hypointense bands adjacent to an abnormal angulation of bowel loops should arise the suspect of fibrotic adherences.

Another congenital condition that can lead to decreased peristalsis and intestinal obstruction is Cystic fibrosis (CF).

CF is an autosomal recessive disorder related to a dysfunction of the chloride ion transport through the epithelium that causes viscous luminal secretions.

As well as the other gastrointestinal organs, the small and large bowel can be involved, with the patient presenting abdominal distension and colicky pain. The obstruction typically arises at the level of the ileocecal valve due to accumulation of dehydrated stool with evidence of a palpable fecal mass [10].

Considering the young age of CF patients and their frequent need of undergoing radiological examination for several reasons, MRE must be considered for intestinal assessment in order to spare further and unnecessary radiation exposure [11].

7.2.2 Ulcerative Colitis

Rather than CD, Ulcerative Colitis (UC) is a chronic inflammatory condition that is mainly limited to the mucosal and submucosal layers of the large bowel, with a contiguous distal to proximal progression [12].

Owing to the supramentioned features, assessment of UC is a prerogative of colonoscopy and imaging modalities are usually not essential.

When endoscopy cannot be performed or completed, MR-Colonography can be effectively employed for large bowel assessment. However, this technique requires 1.5–2.5 L of warm water enema solution administrated through a rectal probe, after previous colonic cleansing with laxatives and therefore might result quite uncomfortable for the patient [13].

In noncompliant patients, the lack of invasiveness makes MRE a palatable alternative.

In such cases a sufficient colonic filling is mandatory.

Although slighter, UC findings resemble those of CD: wall thickening with hyperenhancement after Gadolinium injection and water restriction on DWI scans (Fig. 7.2).

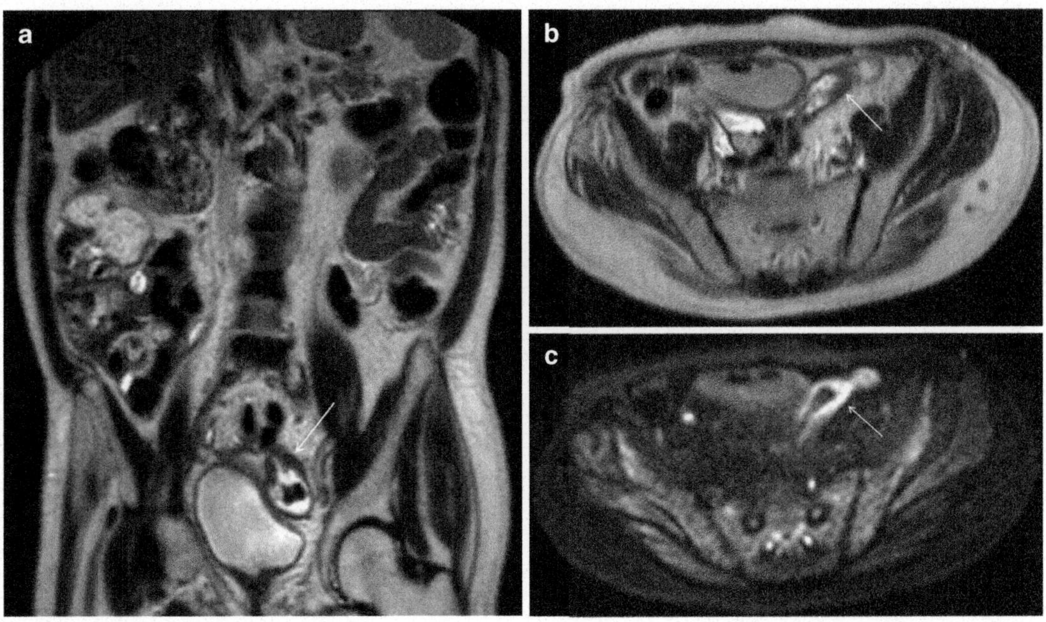

Fig. 7.2 An adequate dilation of large bowel loops was obtained at MRE, thus allowing the detection of a thickening of the sigmoid walls (arrow) on both coronal (**a**) and axial (**b**) T2-w HASTE in a UC patient. The mural thickening is also characterized by high signal intensity (arrow) on axial DWI scan at 800 s/mm² b-value (**c**)

Mucosal edema determining blunting of the colonic haustra is characteristic.

In chronic stages, submucosal fat deposition and fibrosis may also lead to colonic strictures [9, 11, 12].

Penetrating patterns with fistulas formation and fibrofatty proliferation are more prevalent on CD, although they cannot be excluded in UC.

Involvement of the terminal ileum can arise in the form of a primary localization or as a backwash ileitis consequent to reflux of colonic content due to impaired large bowel peristalsis. This localization further complicates the differential diagnosis from CD at radiological Imaging [9, 11, 14].

7.2.3 Appendicitis

Diagnosis of appendicitis results from clinical symptoms and laboratory tests. Ultrasound (US) can help in identifying the appendix with an estimation of wall thickness and mural vascularization.

Lower avalability and prolonged scan times make MRI less suitable than US in case of acute onset.

However, appendicitis has to be considered when performing MRE in patient with lower-right abdominal pain and a diagnosis of CD is not yet established.

Appendicitis features consist of increased caliber of cecal appendix (>7 mm), mural hyperintensity on T2-weighted and DWI scans, mucosal hyperenhancement, and intraluminal filling defects due to appendicolith.

Additional findings may be related to surrounding fat tissue edema, free fluid or abscess, and lymph adenopathies in the right iliac fossa (Fig. 7.3) [2, 3, 9, 11, 15].

7.2.4 Celiac Disease

Celiac disease (CeD) is an autoimmune condition resulting from a combination of genetic and environmental factors that are characterized by malabsorption and chronic intolerance to gluten proteins contained in cereals [16].

The gold standard for CeD diagnosis is endoscopy with jejunal biopsy.

All the radiologic imaging modalities, including barium studies, US, CT, and MRI, can be employed for intestinal assessment of CeD.

Although radiologic findings are non-pathognomonic, they can help in achieving the right diagnosis [17].

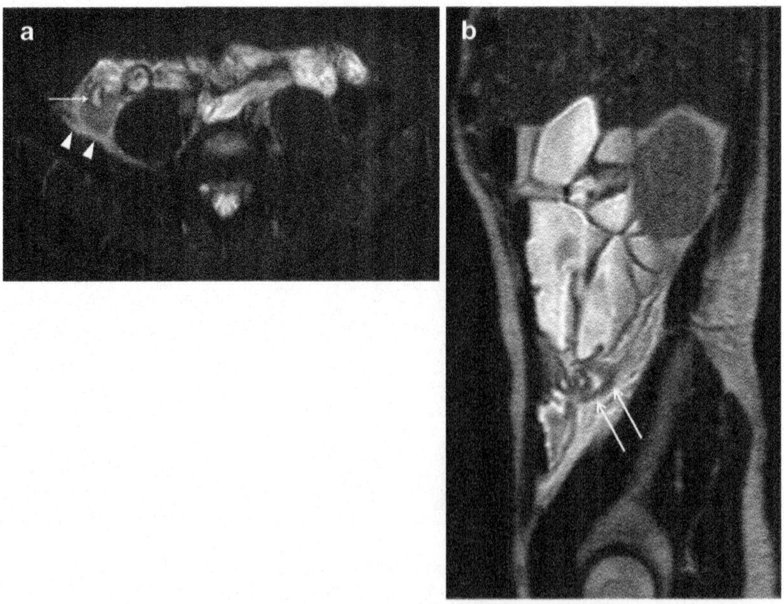

Fig. 7.3 Axial T2-w fat-saturated HASTE (**a**) and sagittal T2-w HASTE (**b**) scans show thickening of the cecal appendix (arrows) due to inflammation. Edema of the surrounding fat tissue (arrowheads) is also visible

MRE appraisal of CeD can show dilation of proximal small bowel loops due to fluid hypersecretion and malabsorption caused by inflammation and mild thickening of walls and folds.

In particular, the latter phenomenon consists of a reversal representation of the jejunoileal folds (the so-called "ileal jejunalization") and has been often referred to as celiac disease, although not specific.

Duodenum and jejunum may demonstrate submucosal fat infiltration as well.

Impaired peristalsis may result in intussusception episodes.

Perivisceral stranding or edema, mesenteric vascular engorgement, lymphadenopathies, and hyposplenism can also be detected [2, 9, 17].

Complications of CeD are represented by ulcerative jejunoileitis, mainly represented by multiple jejunal ulcers, and occurrence of malignancies, such as lymphoma and adenocarcinoma [16].

7.2.5 Infections

Several infections can arise within the gut, including tuberculosis, histoplasmosis, and Whipple disease.

At MRE, infectious enteritis or colitis shows radiologic features similar to those of CD, such as circumferential thickening of bowel walls and folds with contrast hyperenhancement and enlargement of mesenteric lymph nodes [3, 17].

Perforation, fistulas, and abscesses can occur as complications [3].

Intestinal tuberculosis usually involves the ileocecal region and the right hemicolon [18].

Higher suspicion of Whipple disease may arise from detection of nodular-shaped mural thickenings and fat-containing mesenteric lymph nodes [17].

7.2.6 Eosinophilic Gastroenteritis

Eosinophilic gastroenteritis (EGE) is a rare inflammatory disorder that mainly affects the stomach and the proximal segments of the small bowel. According to Klein's classification, three patterns are recognized on the basis of the mural layer involved (mucosal, muscular, and subserosal/serosal).

The final diagnosis is achieved through a combination of clinical picture and endoscopy with biopsy.

Radiological findings are not pathognomonic [19].

The "halo sign," related to a layered appearance due to submucosal edema, and the "araneidlimblike sign," due to crowding of thickened intestinal folds have been mainly associated to EGE and can be considered in the differential diagnosis from malignancies [20].

Thickening of intestinal walls and obstruction can be detected as well (Fig. 7.4) [19].

The serosal/subserosal pattern is usually associated to eosinophil-rich ascites and eosinophilic mesenteric lymphadenopathies [20, 21].

Up to now, the performance of MRE in EGE has been reported only once [11].

Nevertheless, MRE was capable of demonstrating an increased representation of folds and mural thickenings within the jejunal loops. Its use should be encouraged especially in young patients.

7.2.7 Systemic Sclerosis

Systemic sclerosis (also known as "scleroderma") is a chronic autoimmune condition affecting the connective tissue that determines an abnormal accumulation of collagen progressing to fibrosis.

The skin, the vessels, and the internal organs can be involved.

When the gut is affected, esophagus and small bowel represent the most common localizations [10, 17].

Collagen deposition within the intestinal wall causes vascular injury and hypotrophy of smooth muscle resulting in impaired peristalsis.

Therefore, imaging features of systemic sclerosis are consistent with a pseudo-obstruction and include dilated lumen and crowding of the folds with a "hidebound" appearance, due to the decreased separation between the folds themselves. In particular, the latter finding is caused

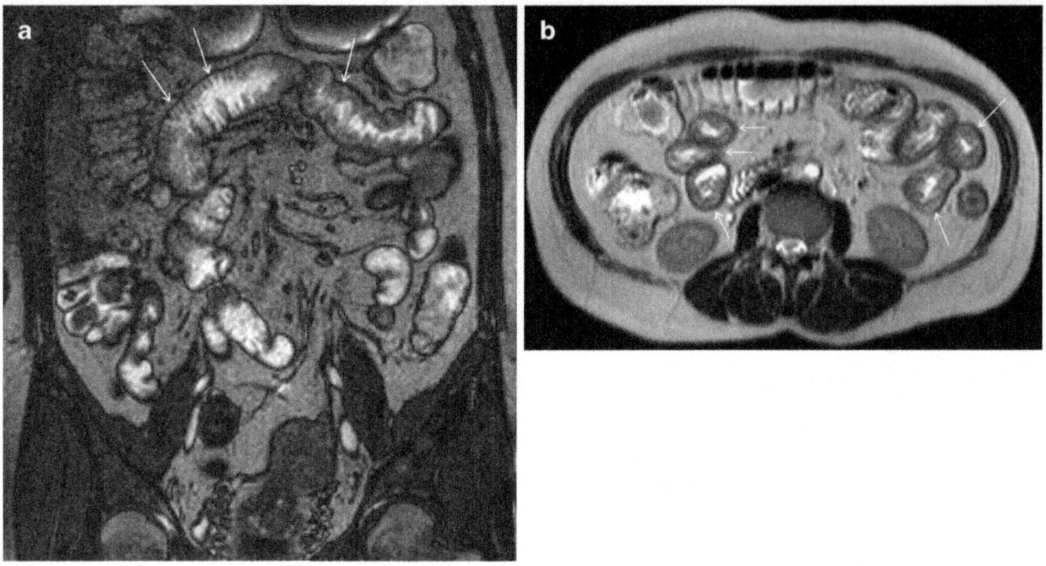

Fig. 7.4 An increased number of folds and mural thickening (arrows) are appreciable within the jejunoileal loops on coronal True-FISP (**a**) and axial HASTE T2-w (**b**) in a patient affected by eosinophilic gastroenteritis

by a peristaltic effort prevented by the fibrotic involution of submucosal and muscolaris propria layers [2, 10, 17].

Additional signs may be represented by sacculations along the antimesenteric edge and pneumatosis cystoides [10].

7.2.8 Amyloidosis

Amyloidosis is a rare genetic or acquired disease caused by deposition of insoluble extracellular protein in different organs [17].

Gastrointestinal involvement, due to deposition within the muscularis mucosae, has been reported in a range varying from 3.3% to 16.8% of patients affected by amyloidosis [22].

Features of intestinal ayloidosis described at CT are non-specific, consisting in wall and fold thickening which can lead to a polypoid appearance or ileal "jejunization" [10, 17].

MRI can further characterize amyloidomas, demonstrating hypointensity on T2-weighted scans due to high protein concentration.

MRE can accurately demonstrate wall and fold thickening, taking advantage of the endoluminal dilation, as well as lymph node infiltration (Fig. 7.5) [8, 23].

7.2.9 Noninfective Vasculitis

The term "vasculitis" encompasses a series of diseases that are generally distinguished on the basis of the size of the vessels involved (large-, medium- and small-vessels vasculitides).

The underlying process is based on inflammation and necrosis of the vascular walls [24].

The vasa recta and intramural arteries within the gastrointestinal tract are frequently involved in patients affected by systemic vasculitis.

Complications are represented by bowel ischemia and intestinal hemorrhage due to aneurysm formation and rupture [24].

CT scan takes advantage of high spatial resolution and the rapid scan time which is of the utmost importance in case of emergency.

On the other hand, MRI can be employed on selected patients, such as young and pregnant.

Furthermore, MRE can provide further details about the intestinal loops in case of ischemia, like thickened walls, scarce contrast enhancement, and obstruction.

However, up to now the performance of MRE has been rarely reported and it has been mainly employed on Systemic lupus erythematosus, with demonstration of wall thickening, indentation of the mucosal and serosal layers due to

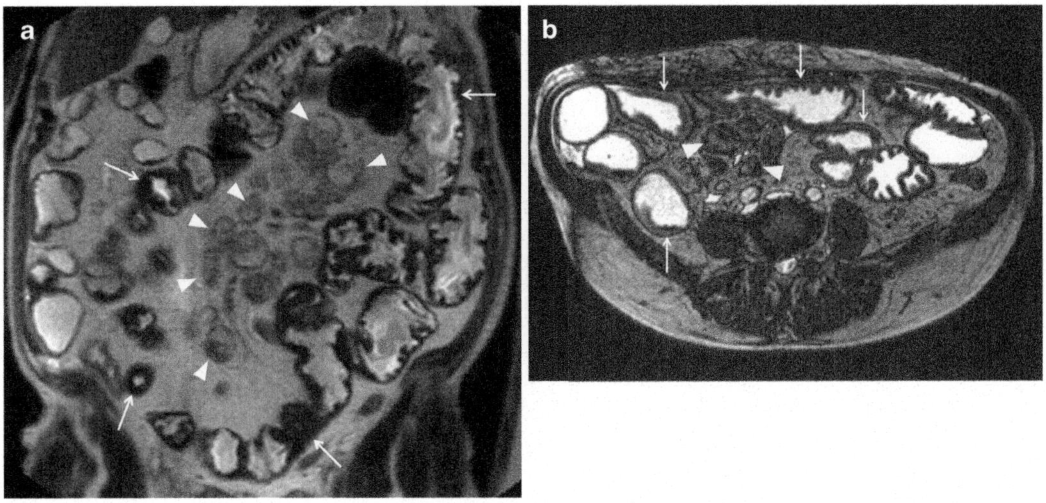

Fig. 7.5 Systemic amyloidosis. On coronal T2-w HASTE (**a**) and axial True-FISP (**b**) a diffuse mural thickening of the small bowel (arrows) with increased number of folds and several mesenteric lymph nodes (arrowheads) are identifiable

edema or hemorrhage of the submucosal layer, and on Behcet disease, in which a homogeneous enhancement and a characteristic polypoid mural thickening were detected [8, 11, 18, 25–27].

7.3 Neoplastic Lesions

The assessment of masses arising within the small bowel is currently a prerogative of CT scan.

However, this evaluation can be challenging within when loops are collapsed or impaired by peristaltic spasms that can mimic a concentric wall thickening.

Although CT Enterography can overcome these limitations, MRE can take advantage of both the luminal distention and the possibility of a safe repetition of the scans, due to the lack of ionizing radiation.

Moreover, MR has a higher soft tissue resolution than CT, which can be useful for more characterization of the lesion features [28].

MRE could be employed for initial staging of patients with suspicious intestinal masses as a guide for enteroscopy or colonoscopy or as an alternative technique in symptomatic patients with a not conclusive endoscopy of the upper and lower intestinal tracts [29].

7.3.1 Benign Neoplastic Lesions

7.3.1.1 Polyps

MRE has been demonstrated to be a reliable alternative to capsule endoscopy (CE) for evaluation of large, small bowel polyps (>15 mm) [30].

Rather than CE, MRE can better localize these lesions and detect extra-luminal malignancies [26, 30].

Polyps can have a sessile (flat) or a pedunculated shape [28].

When the size is lower than 1 cm, they are considered not significant from the clinical point of view [30].

MRE features include low signal on B-FFE sequences, avid contrast enhancement and water restriction on DWI scans (Fig. 7.6) [3, 9, 28].

If cystic, hyperintensity on T2-w SSH scans can be appreciated [3].

On TSE acquisitions, pedunculated polyps must be differentiated from luminal flow artifacts [9].

The absence of radiation exposure makes MRE particularly suitable for follow-up assessment of inherited polyposis syndromes (neurofibromatosis, Peutz–Jeghers polyposis, Gardner syndrome, juvenile polyposis syndrome) within the pediatric population in order to rule out a malignant degeneration [3].

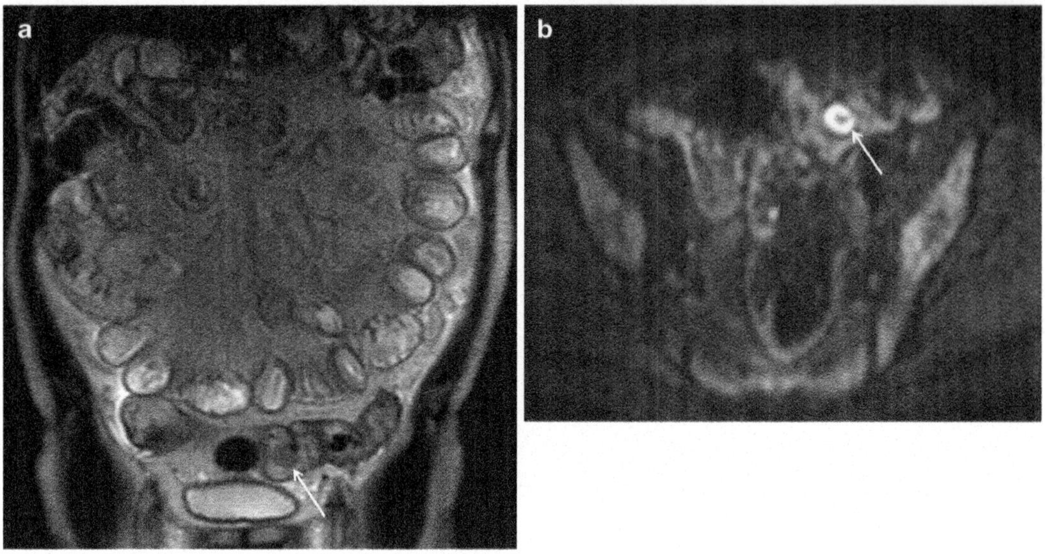

Fig. 7.6 Coronal T2-w HASTE scan (**a**) shows a slightly hypointense round lesion within the sigmoid lumen (arrow) characterized by high-intensity signal on DWI at 800 s/mm^2 b-value (**b**). The endoscopy confirmed the presence of a sigmoid polyp

7.3.1.2 Adenomas

Among the small bowel benign tumors, adenomas are the most common type [28].

These lesions arise from glandular epithelium, are usually of small size (<2 cm) and can have a pedunculated, a sessile, or intramural nodular appearance [28, 31].

Their typical localization is the second portion of the duodenum, whereas they are less common in the ileum.

Small bowel adenomas are characterized by hypointensity on HASTE and B-FFE with homogeneous enhancement after Gd injection.

MR fluoroscopy technique can be helpful in demonstrating an intraluminal filling defect that is generally not associated to an upstream lumen dilation [28].

Considering the risk of malignant transformation or obstruction (also due intussusception), lesions exceeding 1 cm are generally removed by surgery or endoscopy [31].

7.3.1.3 Leiomyomas

Leiomyoma is a rare mesenchymal tumor [31].

When occurring within the small bowel, the jejunum is the usual localization [28]. It can be detected as a round mass with homogeneous contrast hyperenhancement and a mural or intraluminal growth, both of which can be accurately demonstrated by fluoroscopic acquisitions [28, 31].

7.3.1.4 Lipomas

Lipomas are fat-containing lesions that take origin from the submucosal layer [28].

The typical small bowel localizations are, by frequency, the ileum and the duodenum [28, 32].

Lipomas are generally appreciated as sessile, well-circumscribed lesions with intramural or intraluminal development.

MRI features comprise hyperintensity on T1-weighted scans and isointensity to the mesenteric adipose tissue on T2-weighted images. Drop of signal intensity can be demonstrated using fat-saturated acquisitions.

Commonly, intralesional contrast enhancement is not appreciable [28, 32].

7.3.2 Malignant Neoplastic Lesions

7.3.2.1 Gastrointestinal Stromal Tumors

Gastrointestinal stromal tumors (GISTs) represent the most frequent mesenchymal lesions of the gut [28, 33].

GISTs are mainly located in the stomach, whereas only a lower percentage (20–30%) can be identified within the small bowel or in the anorectal region (7%) [28].

GISTs usually appear as large exophytic masses with a dislocating effect on the adjacent intestinal loops, although they can also assume a polypoid shape [2, 28].

They are usually hyperintense on T2-weighted scans with a moderate to marked contrast enhancement [28, 34].

When large in size, foci of hemorrhage, necrosis, or cystic degeneration can be found [28].

MRE can be exploited to better characterize the lesion structure and intraluminal growth.

DWI scans included in the standard MRE protocol may be also useful during follow-up assessment, since an increase in ADC values can be ascribed to a positive treatment response [33].

Lymphnodal spread is not common and distant metastases can be rarely encountered [2].

7.3.2.2 Adenocarcinoma

Although adenocarcinoma is the most common cancer arising in the small bowel (40%), it represents 2% of the whole gastrointestinal tumors [28, 34].

The most common appearance is that of an eccentric circumferential lesion that infiltrates the adjacent loops as well as the perivisceral adipose tissue. Less frequently it can be detected as a polypoid mass.

MRI findings include hypointensity on T2-weighted scans and heterogeneous moderate contrast enhancement (Fig. 7.7).

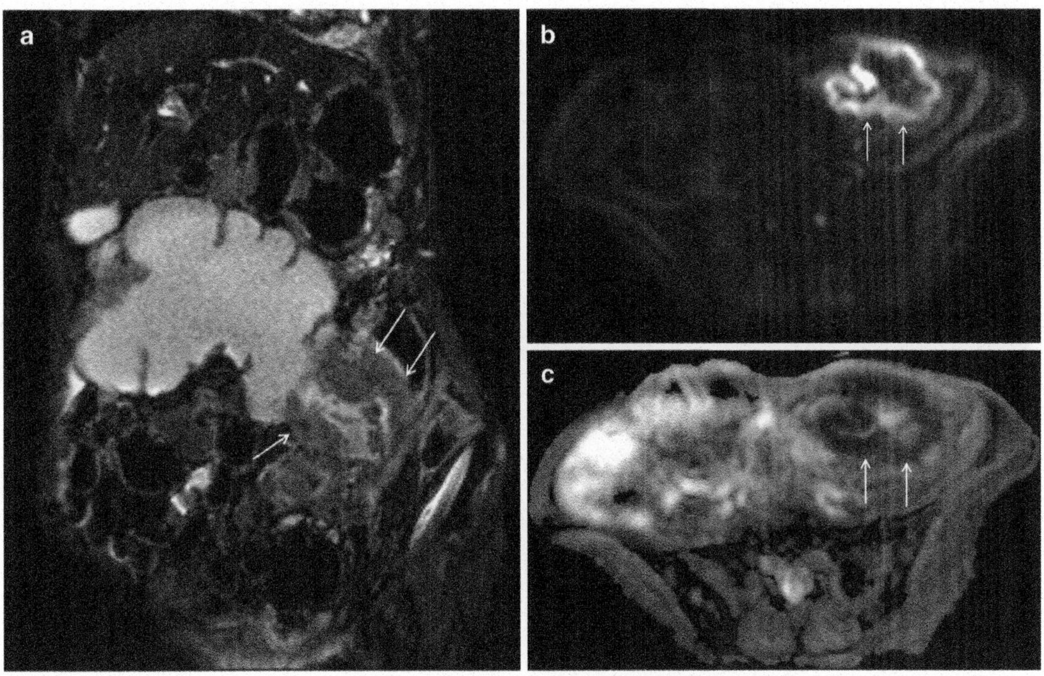

Fig. 7.7 Adenocarcinoma of the left transverse colon. Coronal True-FISP (**a**) shows a huge concentric thickening of the bowel walls (arrows) characterized by hyperintensity on axial DWI scan at 800 s/mm^2 b-value (**b**) and low-intensity signal on the relative ADC map (**c**)

MRE is capable of demonstrating a luminal narrowing, persistent in the different acquisitions of the exam, with upstream fluid dilation [28].

The differential diagnosis of adenocarcinomas includes inflammatory thickening in CD, also considering that these patients are at higher risk for this specific tumor [34, 35].

Moreover, while duodenum is the main localization in general population, jejunum and ileum are preferably affected in CD patients, thus further complicating the differentiation between these two clinical conditions [35].

Lack of "comb sign," detection of a single area of wall thickening and presence of mesenteric lymphadenopathies lean toward the malignant behavior of the lesion [28].

Nonetheless, hyperintensity on DWI at b-values higher than the standard ones (i.e., 3000 sec/mm^2) seems to suggest the presence of a tumor [36].

7.3.2.3 Neuroendocrine Tumors

The performance of MRE in detecting Neuroendocrine tumors (NETs) of the small bowel has shown excellent results in terms of sensitivity [37].

Mural thickening with luminal narrowing, kinking of the involved loops, intense contrast enhancement, and mesenteric lymphadenopathies are typical features of small bowel NETs.

Distant metastases (i.e., hepatic) may also be present [37].

Carcinoid represents a well-differentiated sub-type of NET that takes origin from the enterochromaffin cells.

The distal ileum is the second most common localization after the cecal appendix [28, 34].

Carcinoids are generally hypointense on T1-weighted scans, heterogeneously hyperintense on T2-weighted images and show early and heterogeneous arterial enhancement with delayed washout (Fig. 7.8) [28].

However, it has been demonstrated that in one-third of the cases the primary tumor may be not identifiable on conventional MRI, whereas sensitivity noteworthy increases (93%) using an enteric contrast agent [34, 38].

MR-Enterography or Enteroclysis can be particularly useful in demonstrating the initial mural thickening with luminal protrusion at early stages as well as the final polypoid shape

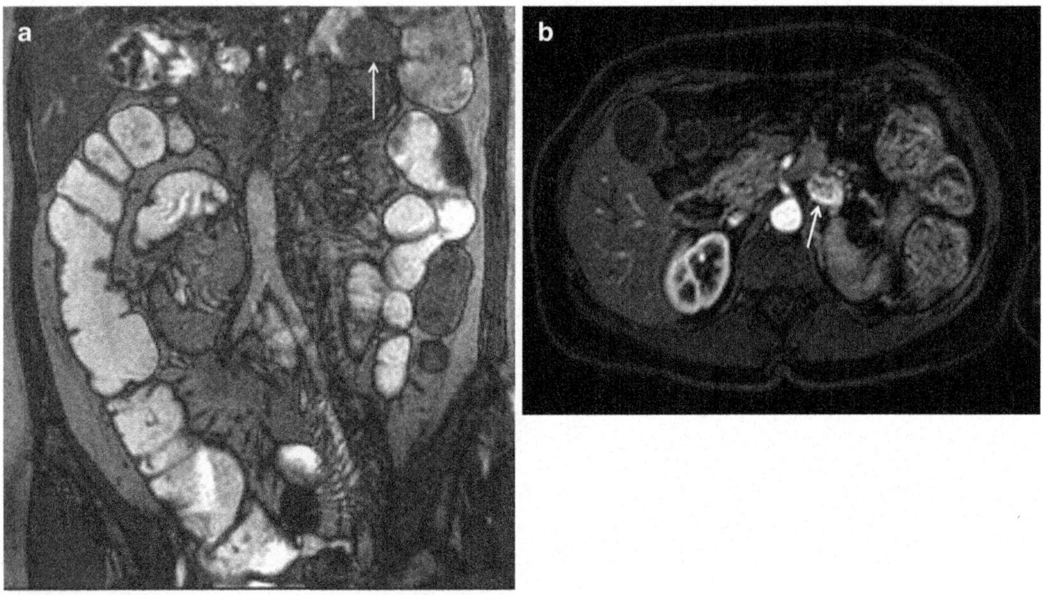

Fig. 7.8 A nodular lesion of approximately 3 cm within the jejunal lumen (arrow) is detectable on coronal True-FISP scan (**a**). The lesion is characterized by avid enhancement in the arterial phase appreciable on axial fat-saturated T1-w GE scan (**b**). The final diagnosis was neuroendocrine tumor

that leads to the formation of intraluminal filling defects [28, 34].

7.3.2.4 Lymphoma

Lymphoma is the third most common neoplasm of all malignant small bowel tumors, accounting approximately for 20% of all the primary small bowel malignancies [2, 28, 34].

The preferred localization is represented by the ileum (>60%), while the most frequent histology is non-Hodgkin B-cell. The T-cell type is usually associated to coeliac disease [2].

Intestinal lymphoma can be detected as a mural thickening, a polypoid lesions with intraluminal protrusion or, although rarely, as a large exophytic mass [2, 28].

Mesenteric adenopathies and splenomegaly can be concomitant [2].

Typical findings of lymphoma at MRE are mucosal ulcerations, wall thickening with luminal narrowing and loss of mucosal folds [34]. Mural contrast enhancement is generally mild [35].

References

1. Mazziotti S, Ascenti G, Scribano E, Gaeta M, Pandolfo A, Bombaci F, Donato R, Fries W, Blandino A. Guide to magnetic resonance in Crohn's disease: from common findings to the more rare complicances. Inflamm Bowel Dis. 2011;17(5):1209–22. https://doi.org/10.1002/ibd.21548. Epub 2010 Nov 5
2. Griffin N, Westerland O. The role of magnetic resonance enterography in the evaluation of non-Crohn's pathologies. Semin Ultrasound CT MR. 2016;37(4):292–300. https://doi.org/10.1053/j.sult.2016.02.002. Epub 2016 Feb 16
3. Judit Machnitz A, Reid JR, Acord MR, Khwaja AB, Biko DM, Ayyala RS, Anupindi SA. MRI of the bowel—beyond inflammatory bowel disease. Pediatr Radiol. 2018;48(9):1280–90. https://doi.org/10.1007/s00247-018-4166-0. Epub 2018 Aug 4
4. Hegde S, Dillman JR, Gadepalli S, Rabah R, Ladino-Torres MF. MR enterography of perforated acute Meckel diverticulitis. Pediatr Radiol. 2012;42(2):257–62. https://doi.org/10.1007/s00247-011-2116-1. Epub 2011 May 19
5. Hammer MR, Podberesky DJ, Dillman JR. Multidetector computed tomographic and magnetic resonance enterography in children: state of the art. Radiol Clin N Am. 2013;51(4):615–36. https://doi.org/10.1016/j.rcl.2013.04.001.
6. Plumb AA, Pendsé DA, McCartney S, Punwani S, Halligan S, Taylor SA. Lymphoid nodular hyperplasia of the terminal ileum can mimic active Crohn disease on MR enterography. AJR Am J Roentgenol. 2014;203(4):W400–7. https://doi.org/10.2214/AJR.13.12055.
7. Cicero G, Pallio S, D'Angelo T, Mazziotti S. Lipomatosis of the ileocecal valve: a not to miss diagnosis when performing magnetic resonance enterography. Clin Case Rep. 2021;9(5):e04316. https://doi.org/10.1002/ccr3.4316. PMID: 34084526; PMCID: PMC8142311
8. Cicero G, Ascenti G, Bottari A, Catanzariti F, Blandino A, Mazziotti S. MR enterography: what is next after Crohn's disease? Jpn J Radiol. 2019;37(7):511–7. https://doi.org/10.1007/s11604-019-00838-y. Epub 2019 Apr 9
9. Anupindi SA, Terreblanche O, Courtier J. Magnetic resonance enterography: inflammatory bowel disease and beyond. Magn Reson Imaging Clin N Am. 2013;21(4):731–50. https://doi.org/10.1016/j.mric.2013.05.002. Epub 2013 Aug 13
10. Hamet B, Durot C, Djelouah M, Adlani I, Marchal A, Arrivé L, Hoeffel C. Involvement of small bowel in systemic disease: CT and MR imaging finding. Clin Imaging. 2020;67:74–85. https://doi.org/10.1016/j.clinimag.2020.05.031. Epub 2020 May 30
11. Cicero G, Blandino A, D'Angelo T, Booz C, Vogl TJ, Ascenti G, Mazziotti S. Mimicking conditions of intestinal Crohn's disease: magnetic resonance enterography findings. Jpn J Radiol. 2022;40(1):19–28. https://doi.org/10.1007/s11604-021-01177-7. Epub 2021 Jul 25
12. Kaushal P, Somwaru AS, Charabaty A, Levy AD. MR Enterography of inflammatory bowel disease with endoscopic correlation. Radiographics. 2017;37(1):116–31. https://doi.org/10.1148/rg.2017160064. Epub 2016 Nov 25
13. Rimola J, Rodríguez S, García-Bosch O, Ricart E, Pagès M, Pellisé M, Ayuso C, Panés J. Role of 3.0-T MR colonography in the evaluation of inflammatory bowel disease. Radiographics. 2009;29(3):701–19. https://doi.org/10.1148/rg.293085115.
14. Yamamoto T, Maruyama Y, Umegae S, Matsumoto K, Saniabadi AR. Mucosal inflammation in the terminal ileum of ulcerative colitis patients: endoscopic findings and cytokine profiles. Dig Liver Dis. 2008;40(4):253–9. https://doi.org/10.1016/j.dld.2007.11.020. Epub 2008 Feb 19
15. Kovanlikaya A, Rosenbaum D, Mazumdar M, Dunning A, Brill PW. Visualization of the normal appendix with MR enterography in children. Pediatr Radiol. 2012;42(8):959–64. https://doi.org/10.1007/s00247-012-2377-3. Epub 2012 Mar 21
16. Masselli G, Picarelli A, Gualdi G. Celiac disease: MR enterography and contrast enhanced MRI. Abdom Imaging. 2010;35(4):399–406. https://doi.org/10.1007/s00261-009-9531-x. Epub 2009 May 13
17. Adamo DA, Sheedy SP, Menias CO, Wells ML, Fidler JL. Malabsorption syndromes, vasculitis, and

other uncommon diseases. Magn Reson Imaging Clin N Am. 2020;28(1):55–73. https://doi.org/10.1016/j.mric.2019.09.001. Epub 2019 Nov 1

18. Amzallag-Bellenger E, Oudjit A, Ruiz A, Cadiot G, Soyer PA, Hoeffel CC. Effectiveness of MR enterography for the assessment of small-bowel diseases beyond Crohn disease. Radiographics. 2012;32(5):1423–44. https://doi.org/10.1148/rg.325115088.
19. Abou Rached A, El Hajj W. Eosinophilic gastroenteritis: approach to diagnosis and management. World J Gastrointest Pharmacol Ther. 2016;7(4):513–23. https://doi.org/10.4292/wjgpt.v7.i4.513. PMID: 27867684; PMCID: PMC5095570
20. Anuradha C, Mittal R, Yacob M, Manipadam MT, Kurian S, Eapen A. Eosinophilic disorders of the gastrointestinal tract: imaging features. Diagn Interv Radiol. 2012;18(2):183–8. https://doi.org/10.4261/1305-3825.DIR.4490-11.1. Epub 2011 Sep 27.9485.12783
21. Zheng X, Cheng J, Pan K, Yang K, Wang H, Wu E. Eosinophilic enteritis: CT features. Abdom Imaging. 2008;33(2):191–5. https://doi.org/10.1007/s00261-007-9209-1.
22. Dahiya DS, Kichloo A, Singh J, Albosta M, Wani F. Gastrointestinal amyloidosis: a focused review. World J Gastrointest Endosc. 2021;13(1):1–12. https://doi.org/10.4253/wjge.v13.i1.1. PMID: 33520102; PMCID: PMC7809597
23. Mainenti PP, Segreto S, Mancini M, Rispo A, Cozzolino I, Masone S, Rinaldi CR, Nardone G, Salvatore M. Intestinal amyloidosis: two cases with different patterns of clinical and imaging presentation. World J Gastroenterol. 2010;16(20):2566–70. https://doi.org/10.3748/wjg.v16.i20.2566. PMID: 20503459; PMCID: PMC2877189
24. Ha HK, Lee SH, Rha SE, Kim JH, Byun JY, Lim HK, Chung JW, Kim JG, Kim PN, Lee MG, Auh YH. Radiologic features of vasculitis involving the gastrointestinal tract. Radiographics. 2000;20(3):779–94. https://doi.org/10.1148/radiographics.20.3.g00mc02779.
25. D'Angelo T, Gallizzi R, Romano C, Cicero G, Mazziotti S. Magnetic resonance Enterography findings of intestinal Behçet disease in a child. Case Rep Radiol. 2017;2017:8061648. https://doi.org/10.1155/2017/8061648. Epub 2017 May 24. PMID: 28630777; PMCID: PMC5463125
26. Gupta A, Postgate AJ, Burling D, Ilangovan R, Marshall M, Phillips RK, Clark SK, Fraser CH. A prospective study of MR enterography versus capsule endoscopy for the surveillance of adult patients with Peutz-Jeghers syndrome. AJR Am J Roentgenol. 2010;195(1):108–16. https://doi.org/10.2214/AJR.09.3174.
27. Cicero G, Blandino A, D'Angelo T, Bottari A, Cavallaro M, Ascenti G, Mazziotti S. Magnetic resonance enterography appraisal of lupus enteritis: a case report. Radiol Case Rep. 2018;13(5):915–9. https://doi.org/10.1016/j.radcr.2018.06.008. PMID: 30069281; PMCID: PMC6068334
28. Masselli G, Guida M, Laghi F, Polettini E, Gualdi G. Magnetic resonance of small bowel tumors. Magn Reson Imaging Clin N Am. 2020;28(1):75–88. https://doi.org/10.1016/j.mric.2019.08.005. Epub 2019 Nov 1
29. Amzallag-Bellenger E, Soyer P, Barbe C, Diebold MD, Cadiot G, Hoeffel C. Prospective evaluation of magnetic resonance enterography for the detection of mesenteric small bowel tumours. Eur Radiol. 2013;23(7):1901–10. https://doi.org/10.1007/s00330-013-2800-7. Epub 2013 Mar 12
30. Tomas C, Soyer P, Dohan A, Dray X, Boudiaf M, Hoeffel C. Update on imaging of Peutz-Jeghers syndrome. World J Gastroenterol. 2014;20(31):10864–75. https://doi.org/10.3748/wjg.v20.i31.10864. PMID: 25152588; PMCID: PMC4138465
31. Jasti R, Carucci LR. Small bowel neoplasms: a pictorial review. Radiographics. 2020;40(4):1020–38. https://doi.org/10.1148/rg.2020200011. Epub 2020 Jun 19
32. Barat M, Dohan A, Dautry R, Barral M, Boudiaf M, Hoeffel C, Soyer P. Mass-forming lesions of the duodenum: a pictorial review. Diagn Interv Imaging. 2017;98(10):663–75. https://doi.org/10.1016/j.diii.2017.01.004. Epub 2017 Feb 6
33. Lu J, Zhou Z, Morelli JN, Yu H, Luo Y, Hu X, Li Z, Hu D, Shen Y. A systematic review of technical parameters for MR of the small bowel in non-IBD conditions over the last ten years. Sci Rep. 2019;9(1):14100. https://doi.org/10.1038/s41598-019-50501-9. PMID: 31575890; PMCID: PMC6773732
34. Soyer P, Boudiaf M, Fishman EK, Hoeffel C, Dray X, Manfredi R, Marteau P. Imaging of malignant neoplasms of the mesenteric small bowel: new trends and perspectives. Crit Rev Oncol Hematol. 2011;80(1):10–30. https://doi.org/10.1016/j.critrevonc.2010.09.010. Epub 2010 Oct 28
35. Placé V, Hristova L, Dray X, Lavergne-Slove A, Boudiaf M, Soyer P. Ileal adenocarcinoma in Crohn's disease: magnetic resonance enterography features. Clin Imaging. 2012;36(1):24–8. https://doi.org/10.1016/j.clinimag.2011.03.006.
36. Yu H, Feng C, Wang Z, Li J, Wang Y, Hu X, Li Z, Shen Y, Hu D. Potential of diffusion-weighted imaging in magnetic resonance enterography to identify neoplasms in the ileocecal region: use of ultra-high b-value diffusion-weighted imaging. Oncol Lett. 2019;18(2):1451–7. https://doi.org/10.3892/ol.2019.10441. Epub 2019 Jun 5. PMID: 31423210; PMCID: PMC6607208
37. Dohan A, El Fattach H, Barat M, Guerrache Y, Eveno C, Dautry R, Mulé S, Boudiaf M, Hoeffel C, Soyer P. Neuroendocrine tumors of the small bowel: evaluation with MR-enterography. Clin Imaging. 2016;40(3):541–7. https://doi.org/10.1016/j.clinimag.2015.12.016. Epub 2016 Jan 15
38. Schmid-Tannwald C, Zech CJ, Panteleon A, Sommer WH, Auernhammer C, Herrmann KA. Characteristic imaging features of carcinoid tumors of the small bowel in MR enteroclysis. Radiologe. 2009;49:242–5.

8 MR-Enterography Pitfalls

Giuseppe Cicero, Tommaso D'Angelo, Giorgio Ascenti, and Silvio Mazziotti

Points of strength of MRE are represented by noninvasiveness, lack of ionizing radiation exposure, comprehensive and multiplanar evaluation of the abdominal cavity [1–4].

On the other hand, radiologists have to evaluate a composite protocol, made of a number of images due to the different types of sequences. The protocol could be also doubled when obtained on the two decubiti.

Further variables, related to patient characteristics or intestinal features, can make MRE evaluation even more challenging.

Therefore, the risk of coming across radiological pitfalls is high. Radiologists should be aware of their occurrence as well as the possible solutions.

Basically, the majority of pitfalls can be classified as related to the MRE protocol, to the intestinal dilation degree and peristaltic contractions.

Drawbacks related to postoperative sequelae and differential diagnosis are explained in detail in the proper chapters.

8.1 Sequences-Related Pitfalls

8.1.1 Half-Fourier Acquisition Single-Shot Turbo Spin Echo Imaging (HASTE)

HASTE is an ultrafast sequence that provides high image quality free from breathing, chemical shift, or susceptibility artifacts [5].

However, the long echo-time, that defines the T2-weighting, makes HASTE sequence prone to peristalsis and flow void artifacts that can impair a correct assessment of the intestinal wall and its content.

In this case, the solution can be represented by the oral administration of spasmolytic agents before the exam starts.

Conversely, the comparison with True-FISP acquisitions, not prone to water motion artifacts, can be sufficient to overcome this weakness (Fig. 8.1) [6–8].

G. Cicero (✉) · T. D'Angelo · G. Ascenti
S. Mazziotti
Department of Biomedical Sciences and Morphological and Functional Imaging, University Hospital Messina, Messina, Italy
e-mail: gcicero@unime.it; tdangelo@unime.it; gascenti@unime.it; smazziotti@unime.it

S. Mazziotti et al. (eds.), *MR Enterography*, https://doi.org/10.1007/978-3-031-11930-9_8

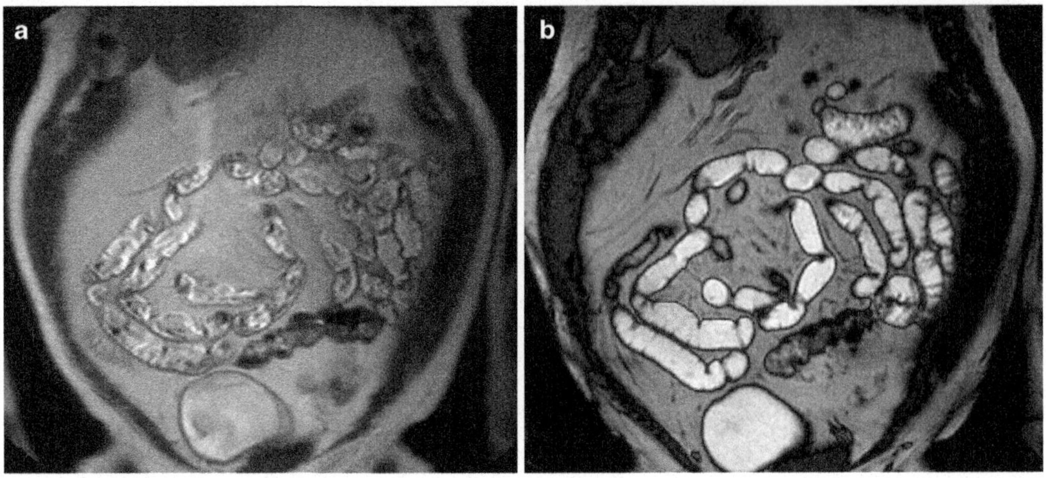

Fig. 8.1 Multiple flow voids, related to the progression of the oral contrast medium, are visible within the ileal loops on coronal T2-w HASTE image (**a**). Due to the shorter acquisition time, coronal B-FFE scan (**b**) is less sensitive to peristalsis and highlights the intensity signal of fluids, thus improving the evaluation of the intraluminal content

8.1.2 Fast Imaging with Steady-State Free Precession

True Fast Imaging with Steady-State Free Precession (True-FISP) removes phase-encoded motion artifacts, thus enhancing fluids [9].

Equally, this sequence can be impaired by susceptibility artifacts (metal or gas related), banding artifacts due to peripheral inhomogeneous magnetic field, and chemical shift artifacts consisting in a black border effect surrounding the intestinal loops [2, 7].

In particular, the latter could lead to a misinterpretation in the assessment of intestinal thickness, in terms of overestimation or false positive findings (Fig. 8.2).

These drawbacks can be well-adjusted by a simple comparison with HASTE sequences [10].

8.1.3 Diffusion Weighted Imaging

In CD patients, small bowel lesions are generally characterized by water molecule restriction [2].

Nonetheless, Diffusion Weighted Imaging (DWI) is subject to motion and susceptibility artifacts, for instance, due to the presence of metal components or intraluminal gas (Fig. 8.3) [11].

Moreover, normal but collapsed bowel loops or neoplastic lesions may show hyperintensity on DWI scans (Fig. 8.4) [12].

In these cases, comparison with the other sequences, change of decubitus, or delayed acquisition with improved intraluminal dilation are advisable.

8.1.4 Ge T1-W FS Pre- and Post-Contrast Medium Injection

Contrast-enhanced phases can be included in a monophasic protocol, with a single acquisition obtained approximately at 70 s after Gadolinium injection (named as "enteric phase"), or comprised within a multiphasic one, consisting of multiple acquisitions usually obtained at 30 s, 60–70 s, 5–7 min [13].

Hyperenhancement of a thickened intestinal wall is considered pathological and can be addressed to both active inflammation and fibrosis.

Hence, false positive findings should be promptly recognized in order to avoid MRE misinterpretations and mistakes in-patient treatment.

In particular, two circumstances have to be acknowledged after Gadolinium injection.

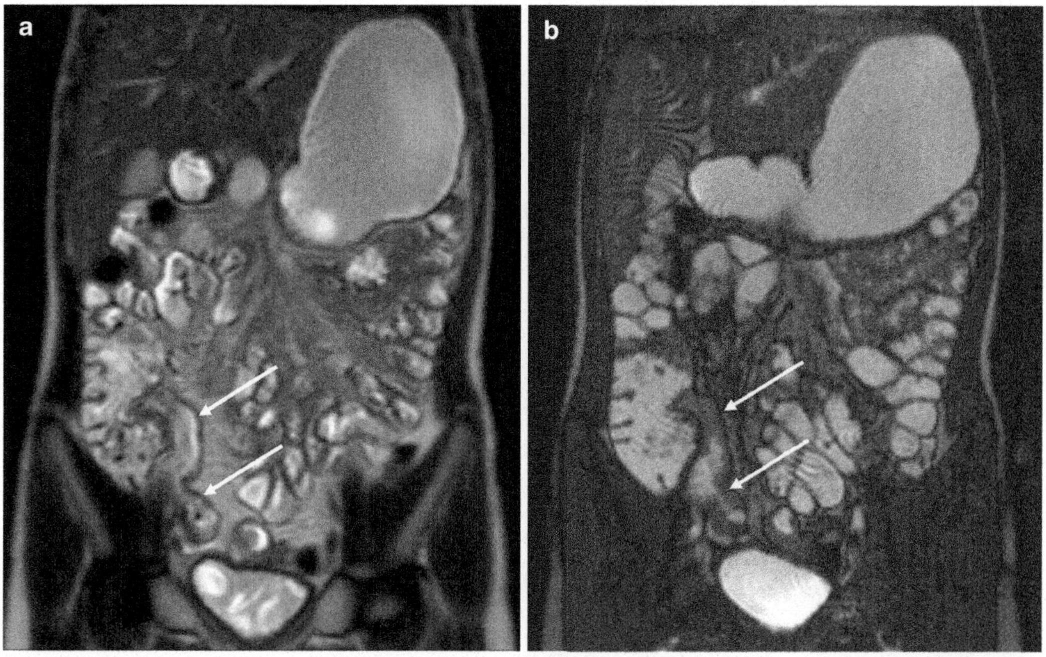

Fig. 8.2 Coronal T2-w HASTE sequence (**a**) shows a thickened last ileal loop (arrows), as well as the coronal B-FFE image (**b**). However, in the latter the wall thickening is magnified by the black border artifact, thus impairing a correct estimation

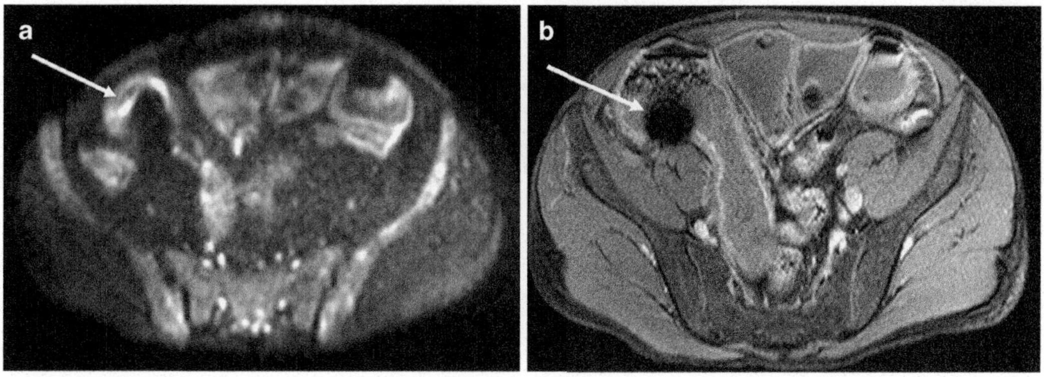

Fig. 8.3 Axial DWI scan at 800 s/mm^2 (**a**) shows high-intensity signal at the level of the last ileal loop (arrow). The comparison with contrast-enhanced axial GE T1-w with FS (**b**) unveiled a magnetic susceptibility artifact (arrow) due to a metal surgical clip placed in the right iliac fossa

First, contrast enhancement can be emphasized whenever bowel loops are collapsed, even if they are normal (Fig. 8.5).

Second, fecal material within the colon is usually characterized by hyperintensity on T1-w sequences with fat saturation. Therefore, it can be mistaken for or can conceal contrast enhancement, respectively, with over- or underestimation of colonic wall lesions.

In this case, subtractive MRI reconstructions, which allow detraction of the precontrast T1-weighted sequence from the enhanced one, can be of great help in identifying a real mural enhancement (Fig. 8.6) [10].

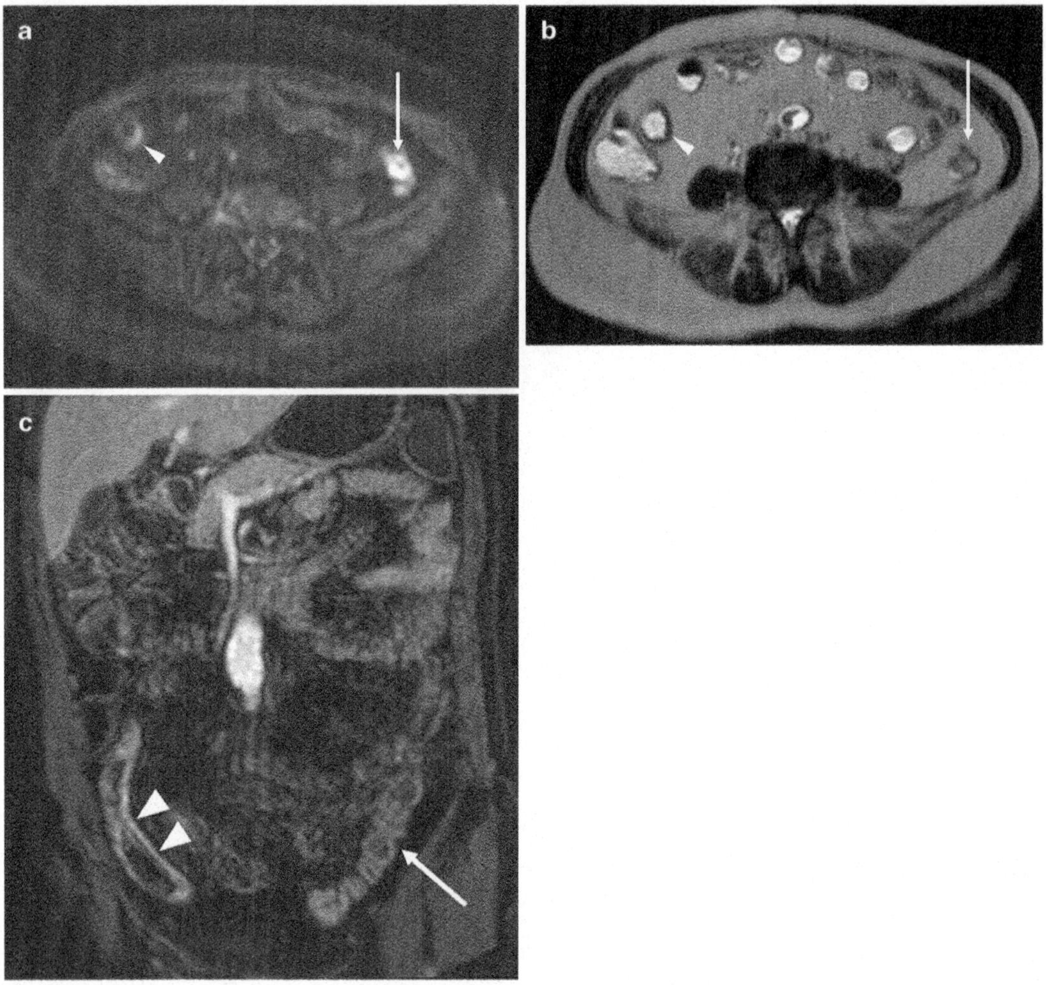

Fig. 8.4 Axial DWI scan at 800 s/mm^2 (**a**) showing hyperintensity on the distal portion of the descending colon (arrow), not confirmed by axial T2-w HASTE (**b**) and coronal contrast-enhanced GE T1-w with FS (**c**). The artifact on DWI is related to a poorly distention with inhomogeneous content due to gasses and fecal material. On the other hand, a real, mild mural thickening is appreciable within the last ileal loop (arrowheads)

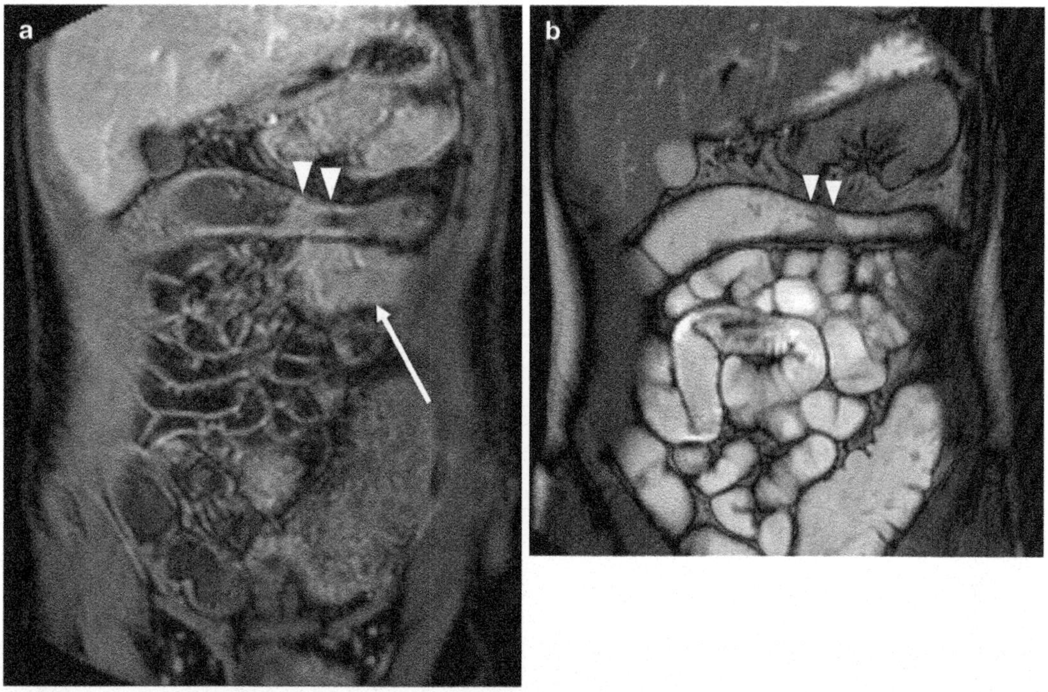

Fig. 8.5 Coronal contrast-enhanced GE T1-w with FS (**a**) shows "pseudohyperenhancement" of jejunal loops (arrow) due to lack of luminal dilation, since no mural thickening is detectable on coronal True-FISP (**b**). A real wall thickening is instead one appreciable within the transverse colon (arrowheads)

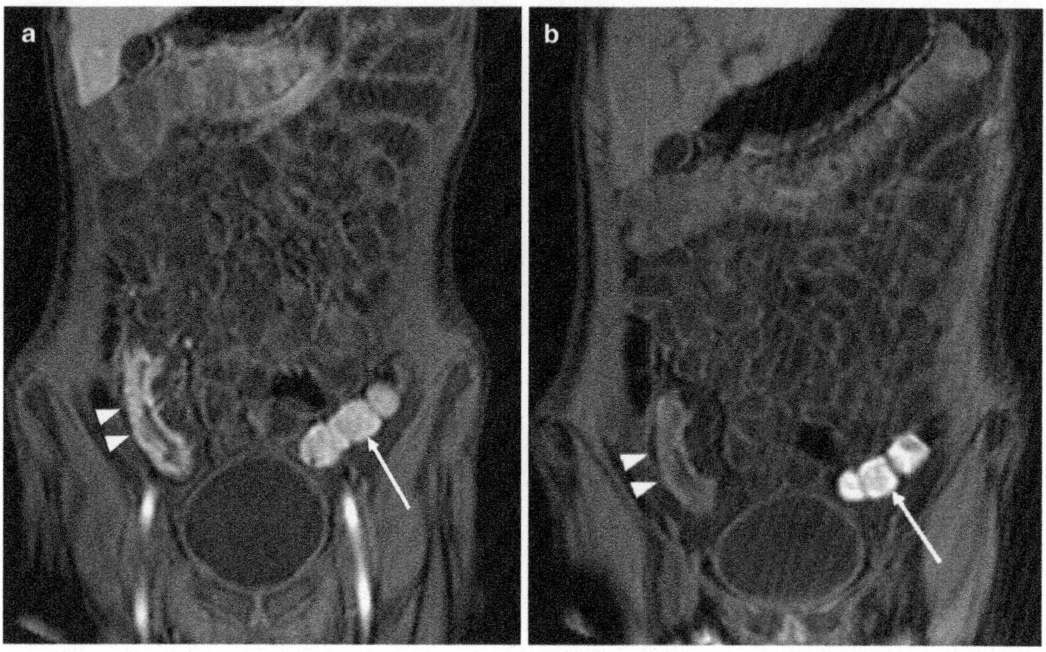

Fig. 8.6 Coronal contrast-enhanced GE T1-w with FS (**a**) showing high-intensity signal on both the ileal (arrowheads) and the sigmoid (arrow) loops, respectively due to real wall thickening and presence of fecal material. A correct identification of mural lesions or assessment of real contrast enhancement requires comparison with unenhanced scans (**b**)

8.2 Intestinal Filling

Suboptimal small-bowel distention is the main cause of under- or overestimation of bowel wall lesions (Fig. 8.7) [10].

In particular, jejunal loops are typically poorly dilated by the oral contrast medium, since they are chharacterized by fast peristalsis.

As mentioned above, collapsed loops generally show a "pseudo-hyperenhancement" after Gd injection and false positive restriction of water molecules on DWI scans.

On the other hand, lack of distention may mask a real wall thickening, leading to a false-negative result.

Patient's intolerance to drinking the full amount of oral contrast medium, with possible occurrence of nausea and vomit, loss of prior fasting, and presence of strictures in the upper intestinal tract are the most common causes of an insufficient intraluminal dilation [8].

MR-Enteroclysis, despite its invasiveness, implies a better distention of jejunal loops and overcome the oral assumption of the enteric agent, which can be problematic, especially in pediatric patients [14].

An alternative approach for the upper midgut appraisal can be achieved through the so-called "MR-Fluoroscopy" technique, which requires an ongoing acquisition of fast sequences during the assumption of the oral contrast agent by the patient (Fig. 8.8) [13].

Cine-MR sequences aid in assessing intestinal peristalsis and in detecting strictures.

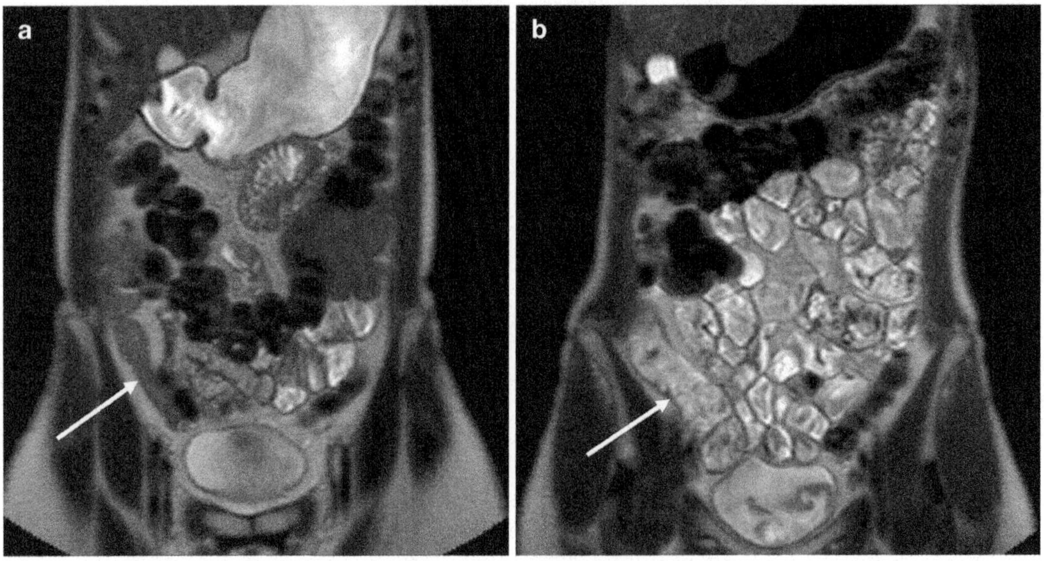

Fig. 8.7 Suboptimal distention of intestinal loops in a CD patient with a previous partial ileocolectomy and ileocolonic anastomosis. On coronal T2-w HASTE image (**a**) a wall thickening was suspected within the right iliac fossa (arrow). The repetition of the same scan after several minutes showed luminal filling and no evidence of mural thickening (**b**)

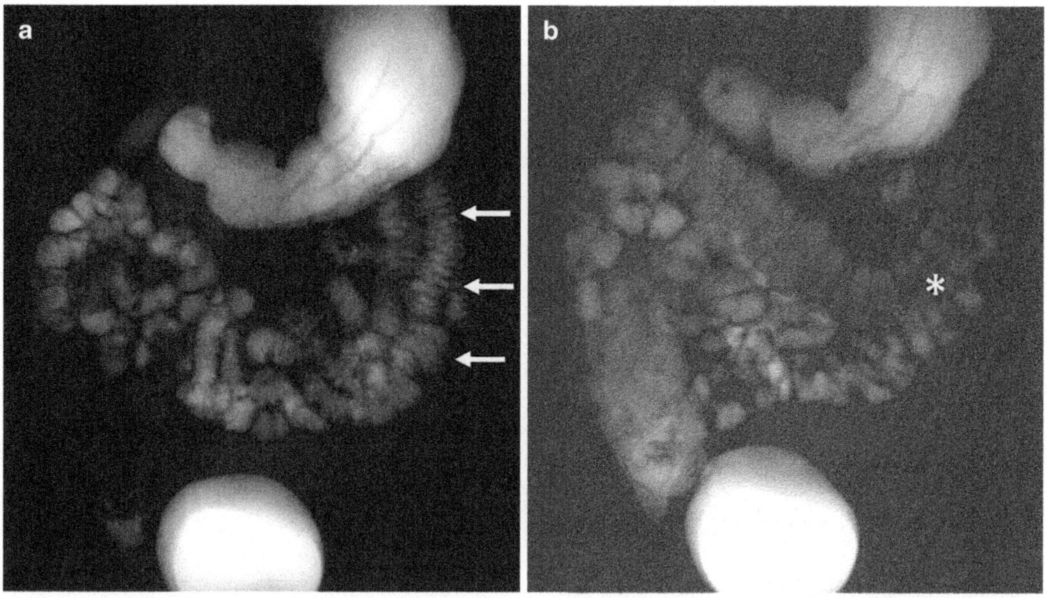

Fig. 8.8 Performance of fast sequences such as coronal T2-w RARE thick-slab (**a**), obtained while the patient is assuming the oral contrast medium inside the scanner, allows following the progressive distension of duodenal and jejunal loops (arrows). The same sequence (**b**), acquired at the standard MRE timeline, demonstrates collapsed jejunal loops and superimposition of transverse colon (asterisk)

8.3 Peristaltic Spasms

Although distended by the oral contrast medium, some loops can appear thickened due to a transient twitching due to physiologic peristaltic movements (Fig. 8.9).

This problem is particularly relevant when MRE is performed without the preliminary administration of spasmolytic agents.

In such a case, comparison with the other sequences, especially True-FISP and Cine-MR, change of decubitus or performance of delayed scans are mandatory in order to avoid mistakes [15].

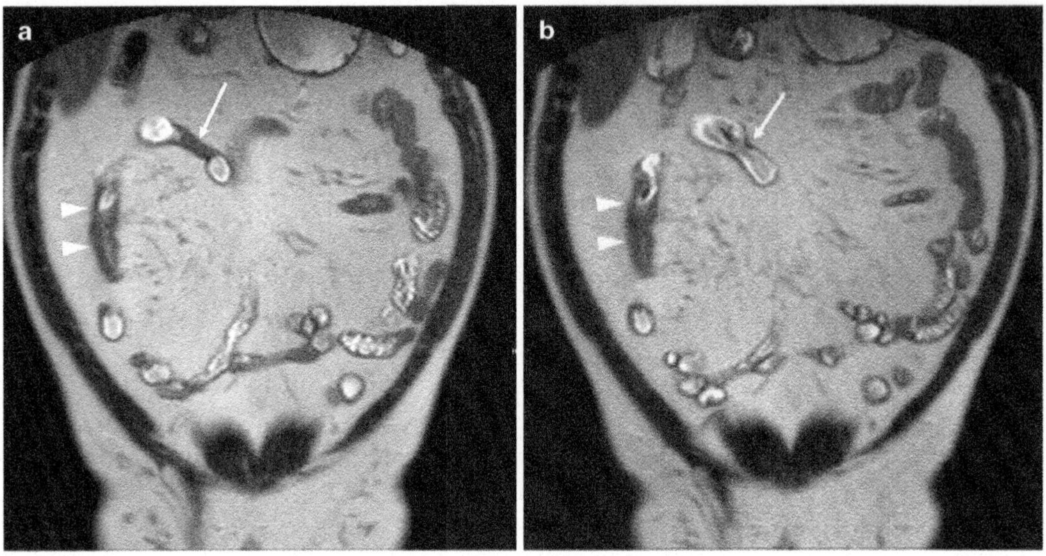

Fig. 8.9 Coronal T2-w HASTE obtained in the supine position (**a**) shows two luminal narrowing suspicious for CD lesions. Only the real wall thickening (arrowheads) is also visible on coronal T2-w HASTE in prone decubitus (**b**), while the other was compatible with a peristaltic spasm (arrow)

References

1. Mazziotti S, Ascenti G, Scribano E, Gaeta M, Pandolfo A, Bombaci F, Donato R, Fries W, Blandino A. Guide to magnetic resonance in Crohn's disease: from common findings to the more rare complicances. Inflamm Bowel Dis. 2011;17(5):1209–22. https://doi.org/10.1002/ibd.21548. Epub 2010 Nov 5
2. Griffin N, Grant LA, Anderson S, Irving P, Sanderson J. Small bowel MR enterography: problem solving in Crohn's disease. Insights Imaging. 2012;3(3):251–63.
3. Allen BC, Leyendecker JR. MR enterography for assessment and management of small bowel Crohn disease. Radiol Clin N Am. 2014;52(4):799–810.
4. Semelka RC, Kelekis NL, Thomasson D, Brown MA, Laub GA. HASTE MR imaging: description of technique and preliminary results in the abdomen. J Magn Reson Imaging. 1996;6(4):698–9.
5. Cicero G, Mazziotti S. Crohn's disease at radiological imaging: focus on techniques and intestinal tract. Intest Res. 2021;19(4):365–78. https://doi.org/10.5217/ir.2020.00097. Epub 2020 Nov 25. PMID: 33232590; PMCID: PMC8566824
6. Sinha R, Verma R, Verma S, Rajesh A. MR enterography of Crohn disease: part 1, rationale, technique, and pitfalls. AJR Am J Roentgenol. 2011;197(1):76–9.
7. O'Malley RB, Hansen NJ, Carnell J, Afzali A, Moshiri M. Update on MR Enterography: potentials and pitfalls. Curr Radiol Rep. 2016;4:42.
8. Tolan DJ, Greenhalgh R, Zealley IA, Halligan S, Taylor SA. MR enterographic manifestations of small bowel Crohn disease. Radiographics. 2010;30(2):367–84.
9. Huang TY, Huang IJ, Chen CY, Scheffler K, Chung HW, Cheng HC. Are TrueFISP images T2/T1-weighted? Magn Reson Med. 2002;48(4):684–8.
10. Ram R, Sarver D, Pandey T, Guidry CL, Jambhekar KR. Magnetic resonance enterography: a stepwise interpretation approach and role of imaging in management of adult Crohn's disease. Indian J Radiol Imaging. 2016;26(2):173–84.
11. Sinha R, Rajiah P, Ramachandran I, Sanders S, Murphy PD. Diffusion-weighted MR imaging of the gastrointestinal tract: technique, indications, and imaging findings. Radiographics. 2013;33(3):655–76. discussion 676-80
12. Chavhan GB, Babyn PS, Walters T. MR enterography in children: principles, technique, and clinical applications. Indian J Radiol Imaging. 2013;23(2):173–8.
13. Lee SM, Kim WS, Choi YH. Pediatric magnetic resonance enterography: focused on Crohn's disease. Pediatr Gastroenterol Hepatol Nutr. 2015;18(3):149–59.
14. Cicero G, D'Angelo T, Bottari A, Costantino G, Visalli C, Racchiusa S, et al. Superior mesenteric artery syndrome in patients with Crohn's disease: a description of 2 cases studied with a novel magnetic resonance enterography (MRE) procedure. Am J Case Rep. 2018;19:431–7.
15. Yu HS, Gupta A, Soto JA, LeBedis C. Emergency abdominal MRI: current uses and trends. Br J Radiol. 2016;89(1061):20150804.

9 MRI of the Anal Region in Crohn's Disease and Beyond

Silvio Mazziotti, Giuseppe Cicero, Alfredo Blandino, and Tommaso D'Angelo

Perianal involvement in Crohn's disease can be complicated by a variety of pathologic conditions [1]. Perianal abscess occurs in up to 80% of patients with CD, as the frequency of perianal fistulas varies from 17% to 43% [2–9].

The risk of perianal involvement increases when the distal gastrointestinal tract is affected. In fact, patients with disease confined to the colon have a higher incidence of perianal fistulas, and the rate can approach 100% in patients with a rectal CD localization [4, 10]. Although the precise etiology of perianal fistulas remains unknown, they may be caused by obstruction of sinus tracts or cryptoglandular infections [1]. The former theory suggests that fistulas may arise due to the extension of deep penetrating ulcers into the anus or distal rectum; over time, fecal material stagnates within the ulcers and the evacuating pressure also facilitates their further extension with formation of a real fistula [1, 11]. According to the latter theory, infection of anal glands gives rise to abscess formation which represents the point of origin of a fistula [12].

Perianal alterations may anticipate the intestinal manifestations of CD by several months or years. About two-thirds of patients with perianal disease will be diagnosed with intestinal disease within 1 year and another third within 1–5 years, with only a few patients being diagnosed after more than 5 years. Only a small proportion of patients with CD may persist in having isolated perianal involvement [13, 14].

Surgical management of perianal CD depends on the extent, localization, and complexity of the fistula. The surgical approach can be more invasive in case of large fistulas and in presence of abscesses.

Therefore, imaging is necessary for surgical planning. The main purpose is to identify the primary fistulous track, especially when a cutaneous opening is absent, the possible secondary tracks, and the sites of any abscess cavities, and to define their anatomic relation with the sphincters, the levator ani muscle, and the ischiorectal and ischioanal fossae [15].

9.1 Classification of Fistulas

Classification systems for perianal fistulizing CD disease are useful in determining what surgical procedure (if any) should be performed. Several classifications have been used to define the different types of fistulas. The most anatomically detailed and used is the one established by Parks. This classification distinguishes four different possible paths: inter-sphincteric, trans-sphincteric, supra-sphincteric, and extra-sphincteric [16]. Since there is no relation

S. Mazziotti (✉) · G. Cicero · A. Blandino
T. D'Angelo
Department of Biomedical Sciences and Morphological and Functional Imaging, University Hospital Messina, Messina, Italy
e-mail: smazziotti@unime.it; gcicero@unime.it; ablandino@unime.it; tdangelo@unime.it

S. Mazziotti et al. (eds.), *MR Enterography*, https://doi.org/10.1007/978-3-031-11930-9_9

between the sphincter or the perianal canal, superficial subcutaneous fistulas are excluded from Parks' classification (Fig. 9.1).

On the other hand, the St. James's University Hospital classification has been developed according to MRI findings. Fistulas are distinguished on a 5-point scale: grade 1, simple linear inter-sphincteric with no ramifications; grade 2, inter-sphincteric with presence of abscess or secondary tract bounded by the external sphincter; grade 3, external sphincter leak with trans-sphincteric path and possible hyperemia or edema arising within the adjacent fossae; grade 4, trans-sphincteric with extension to the ischio-rectal fossa where an abscess can be found; grade 5, the fistulous path extends above the supralevator ani with involvement of the ischio-rectal fossa [17].

Finally, the Standards Practice Task Force of the American Society of Colon and Rectal Surgeons has analyzed this topic from the surgical point of view. This classification merely evaluates the anal disease on the basis of the possible treatments, distinguishing simple fistula-in-ano, complex fistula-in-ano, fistulas associated to CD and abscesses [18].

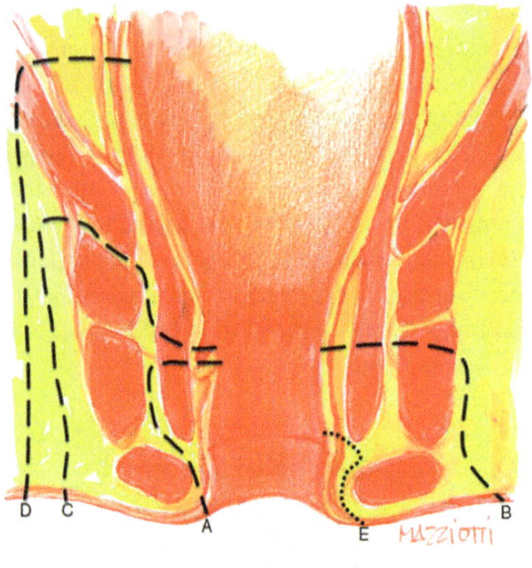

Fig. 9.1 Parks' classification: (A) Inter-sphincteric fistula, (B) Trans-sphincteric fistula, (C) Supra-sphincteric fistula, (D) Extra-sphincteric fistula, (E) Subcutaneous fistula (not part of Parks' classification)

9.2 Technique

MRI of the anal district is performed through a surface phased array coil without any need of lumen dilation.

The field of view must be cranio-caudally extended from the supralevator plane to the anal verge including the surrounding gluteal fat tissue [19–21].

An initial sagittal T2-weighted Turbo-Spin echo (TSE) can be obtained in order to provide an overview of the anal canal and to orientate the following scans.

The following T2-weighted TSE sequences are acquired on oblique-coronal and axial planes respectively orientated perpendicular and parallel to the major axis of the anal canal (Fig. 9.2).

The use of T2-weighted TSE scans with fat saturation significantly facilitates the detection of fluid collection and perilesional edema.

Two- or three-dimensional T1-weighted images with fat saturation obtained before and after the injection of gadolinium contrast material are useful not only for identification of fistulas, but also in distinguishing active inflammation from fibrosis. In fact, enhancement tends to be early and intense in the former condition, whereas it appears more progressive and delayed in the latter.

Finally, oblique-coronal or, preferably, oblique-axial diffusion-weighted images (DWI) highlight the presence of fistulas and abscesses and result particularly important in case of contraindications to intravenous contrast medium injection [19, 20, 22].

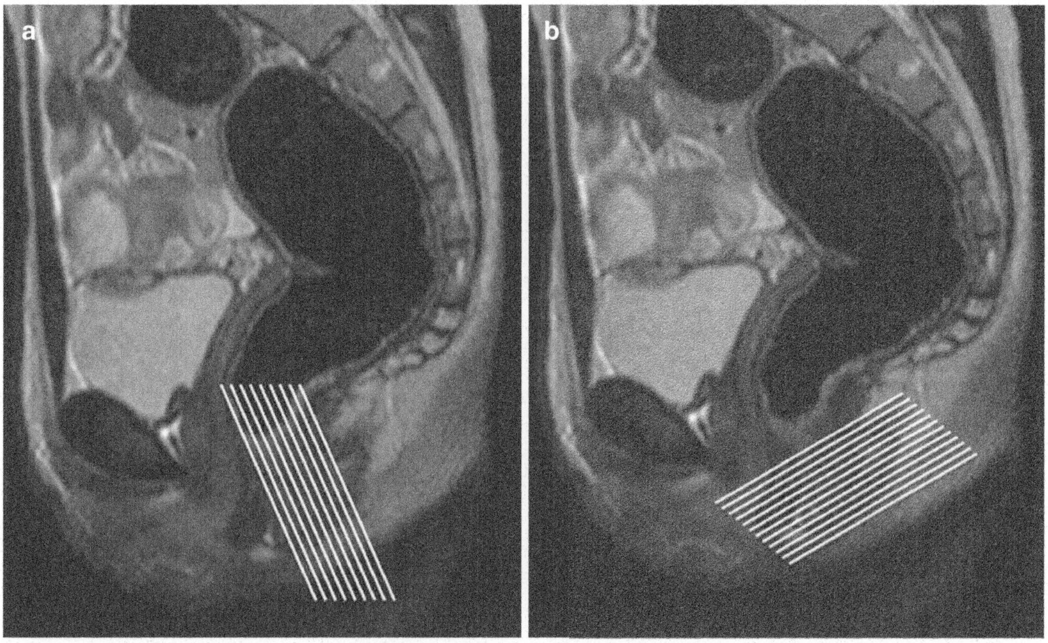

Fig. 9.2 Technique. Orientation of coronal (**a**) and axial (**b**) scan series on a midline sagittal T2-weighted TSE image, respectively, parallel and perpendicular to the major axis of the anal canal

9.3 Perianal Findings in Crohn's Disease

Active fistulous tracts and abscesses filled with pus and granulation tissue appear as hyperintense structures on T2-weighted and STIR images (Figs. 9.3, 9.4, and 9.5). In particular, identification of the tract is easier when evaluating fat-saturated sequences, whereas T2-weighted scans without fat suppression give more detailed information about the anatomic relationship of the tract with the surrounding anatomical structures (Fig. 9.6). Active tracts are often surrounded by hypointense fibrous walls, which can be relatively thick and hypointense, especially in cases of recurrent disease and/or previous surgery (Fig. 9.7). Some hyperintensity in this fibrous tissue may seldom be seen, probably due to associate inflammatory edema. This hyperintensity may also extend beyond the tract and its fibrous sleeve, reflecting adjacent inflammation (Fig. 9.8). On contrast-enhanced T1-weighted images, active fistulous tracks brilliantly enhance, as do the walls of abscess cavities (Figs. 9.3, 9.4, and 9.9). Retained pus remains unenhanced, with a resulting peripheral ring enhancement, a typical appearance of abscess elsewhere in the body.

MRI performed along the coronal and axial planes demonstrates fistulous tracks in relation to the sphincteric complex, ischiorectal and ischioanal fossae, and levator plane. Tracts are described in accordance with the terminology illustrated by Parks et al. [16].

When a fistula is contained by the external sphincter, it is defined as inter-sphincteric (Fig. 9.7). On the contrary, any evidence of a fistulous tract in the ischioanal fossa effectively excludes an inter-sphincteric fistula. Moreover, it should also be considered that trans-, supra-, and extra-sphincteric fistulas share the common feature of a tract that lies beyond the confines of the external sphincter. Although trans-sphincteric fistulas are the most common cause of a tract in the ischioanal fossa (Fig. 9.10), it must be remembered that a differentiation between these three fistulas is possible only by locating the internal

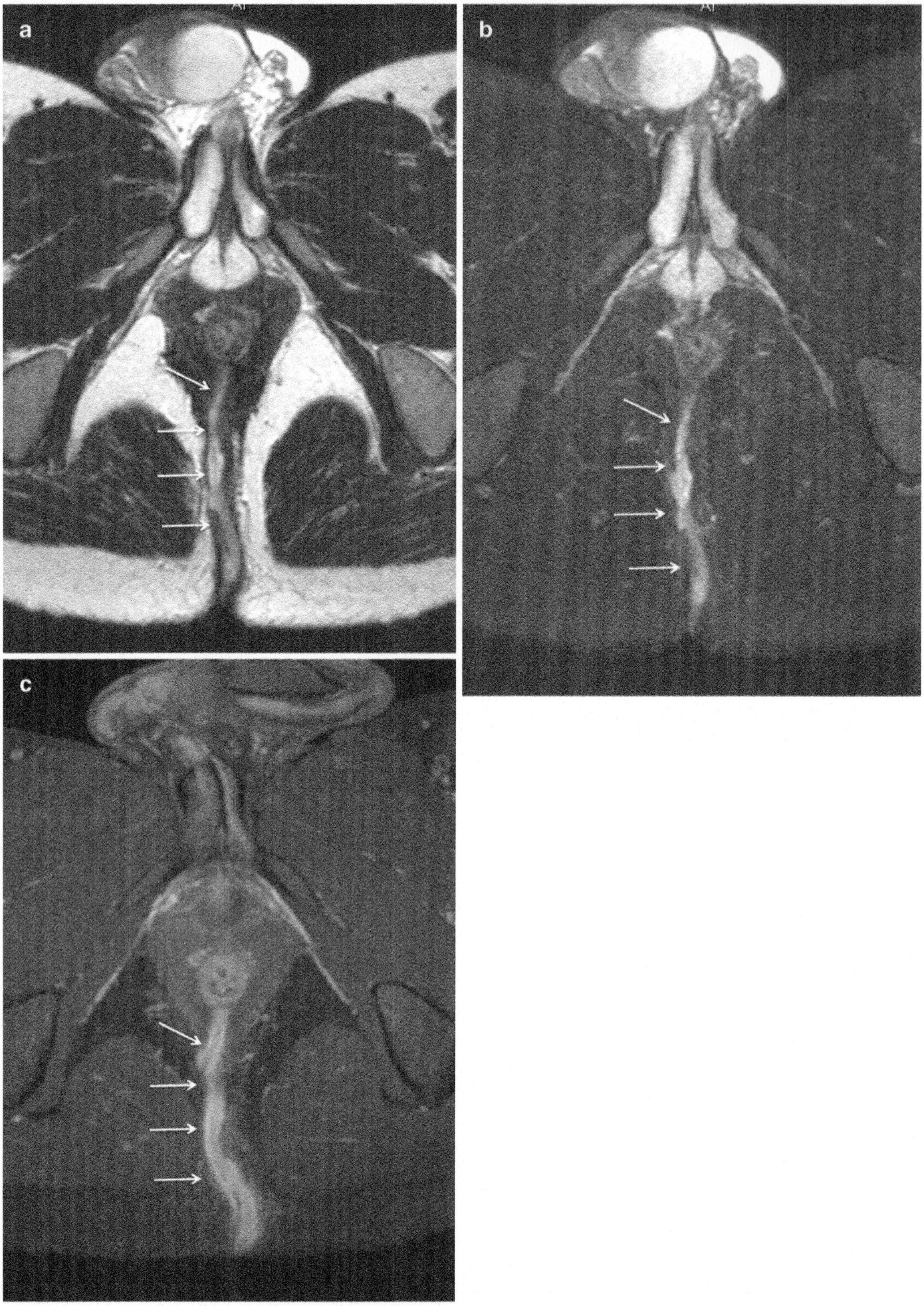

Fig. 9.3 Trans-sphincteric fistula. Axial-oblique T2-weighted TSE images performed without (**a**) and with fat saturation (**b**) show a posterior midline hyperintense trans-sphincteric fistula (arrows). Contrast enhancement of the inflamed wall of the fistula is well depicted after i.v. injection of Gadolinium (**c**)

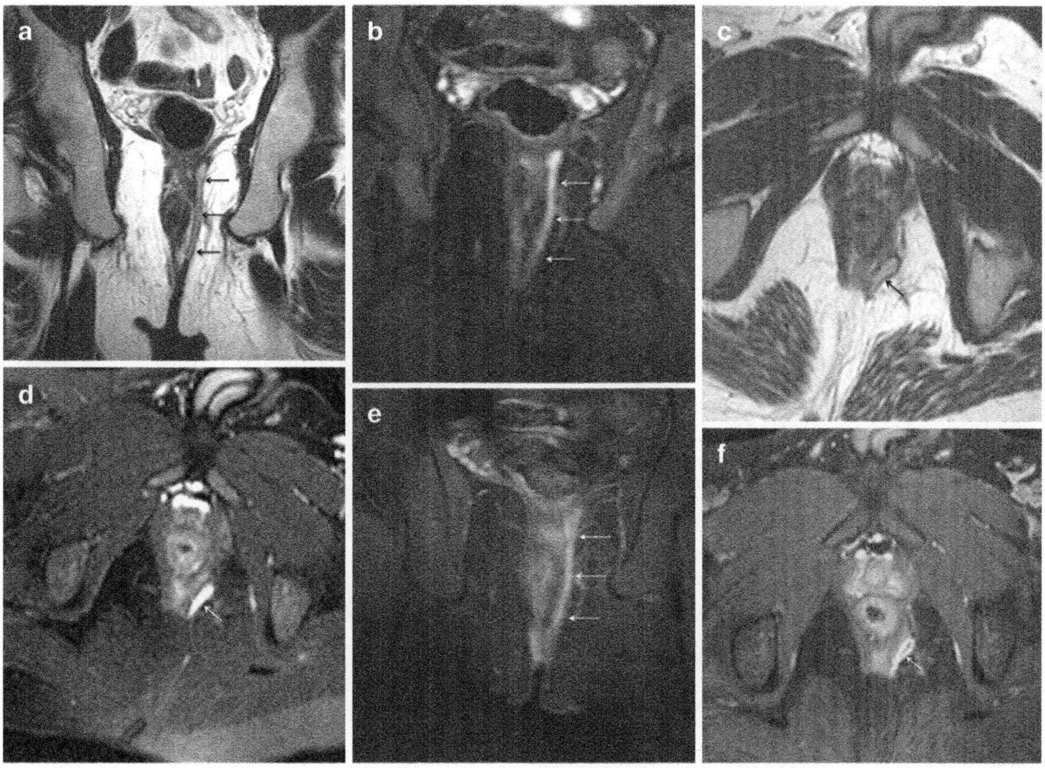

Fig. 9.4 Extra-sphincteric fistula. Coronal-oblique (**a, b**) and axial-oblique (**c, d**) T2-weighted TSE images, performed without (**a, c**) and with fat saturation (**b, d**) show a left hyperintense extra-sphincteric fistula (arrows). On coronal-oblique (**e**) and axial-oblique (**f**) T1-weighted TSE fat-saturated images, performed after Gadolinium administration, enhancement of the fistulous walls is well depicted (arrows)

opening and clearly determining the course between this and the primary tract [23].

Finally, in presence of voluminous perianal abscesses, eventual fistulous tracts can be hidden, due to the compression or dislocation of the sphincter complex itself and, in more severe cases, large dehiscence with nearby pelvic viscera may also be found (Figs. 9.11 and 9.12).

In order to describe the exact site and direction of fistulous tract according to the surgical view, radiologist should correlate the axial MRI findings to the "anal clock."

With this system, key points of the fistula, such as internal opening and path, are referred to as the sphincteric structures using clock benchmarks. Axial sequences are employed for this assessment and position of the patient inside the scanner must be clarified since it can differ from that of clinical examination or surgical intervention [20, 24].

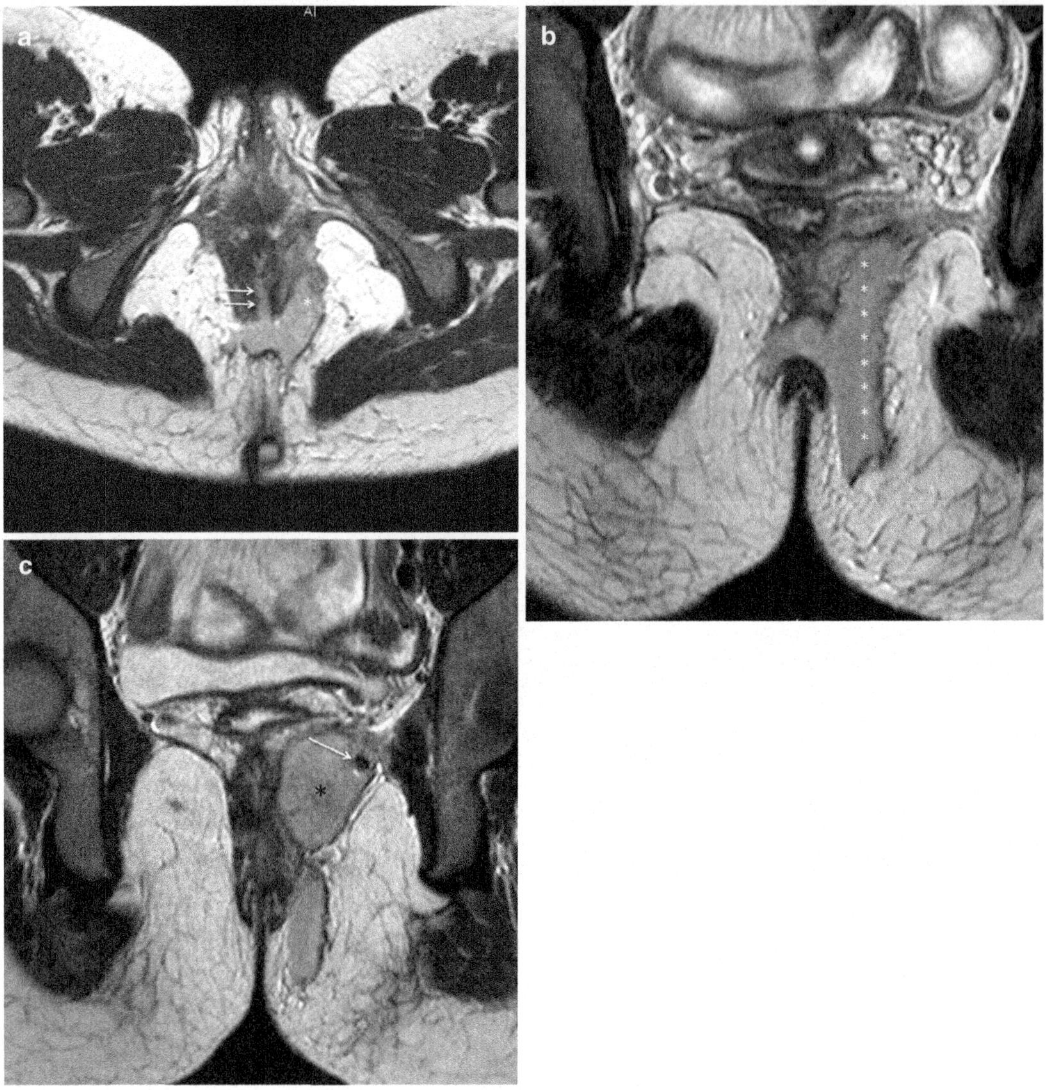

Fig. 9.5 Trans-sphincteric fistula with abscess. Axial-oblique (**a**) and coronal-oblique (**b**, **c**) T2-weighted TSE images show a posterior midline trans-sphincteric fistula (arrows in **a**) with its course in the left ischioanal fossa (white asterisks in a and **b**). A voluminous perianal abscess can also be seen (black asterisk in **c**) with a small gas bubble in its context (arrow)

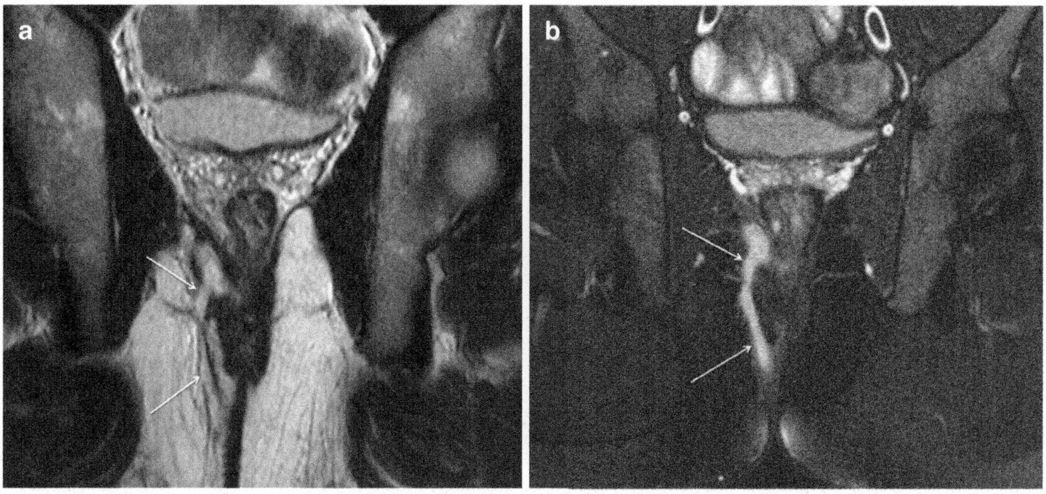

Fig. 9.6 Trans-sphincteric fistula. Coronal-oblique T2-weighted TSE image (**a**) shows a right trans-sphincteric fistula, involving the ipsilateral ischioanal fossa (arrows). Fat-saturated image (**b**) clearly shows the fistulous tract (arrows) due to suppression of fat intensity signal

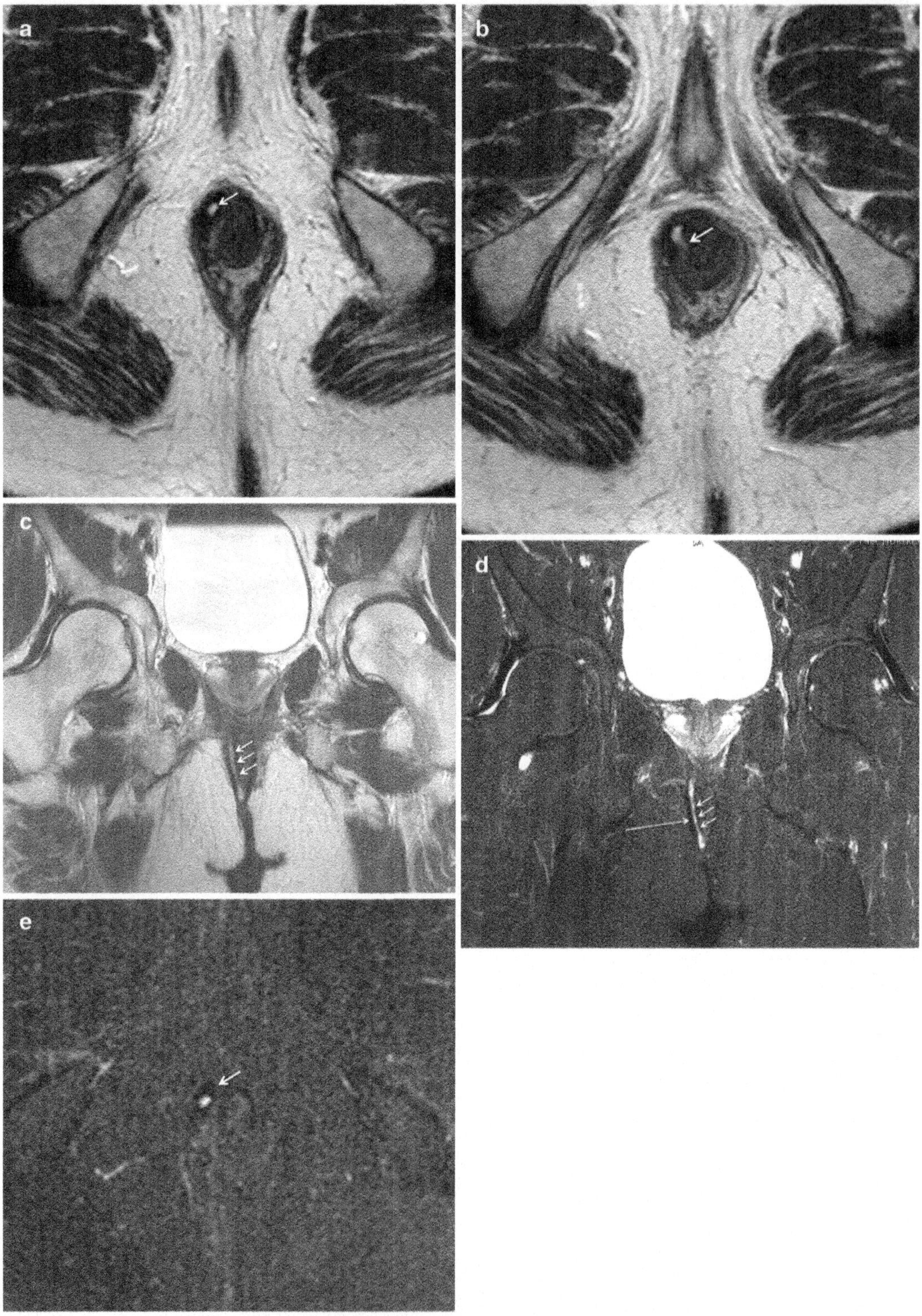

Fig. 9.7 Inter-sphincteric fistula. Axial-oblique T2-weighted TSE images (**a**) show inter-sphincteric fistula at 11 o'clock (arrow). In a more cranial image (**b**) the internal opening is demonstrable (arrow). Coronal-oblique T2-weighted TSE images performed without (**c**) and with fat saturation (**d**) show right inter-sphincteric fistula, traceable along most of its course (short arrows), with sparing of ischioanal fossa. Note the hypointense fibrous wall surrounding the active fistula (long arrow in **d**), also confirmed in the axial T2-weighted TSE fat-saturated image (**e**) obtained through its middle part (arrow)

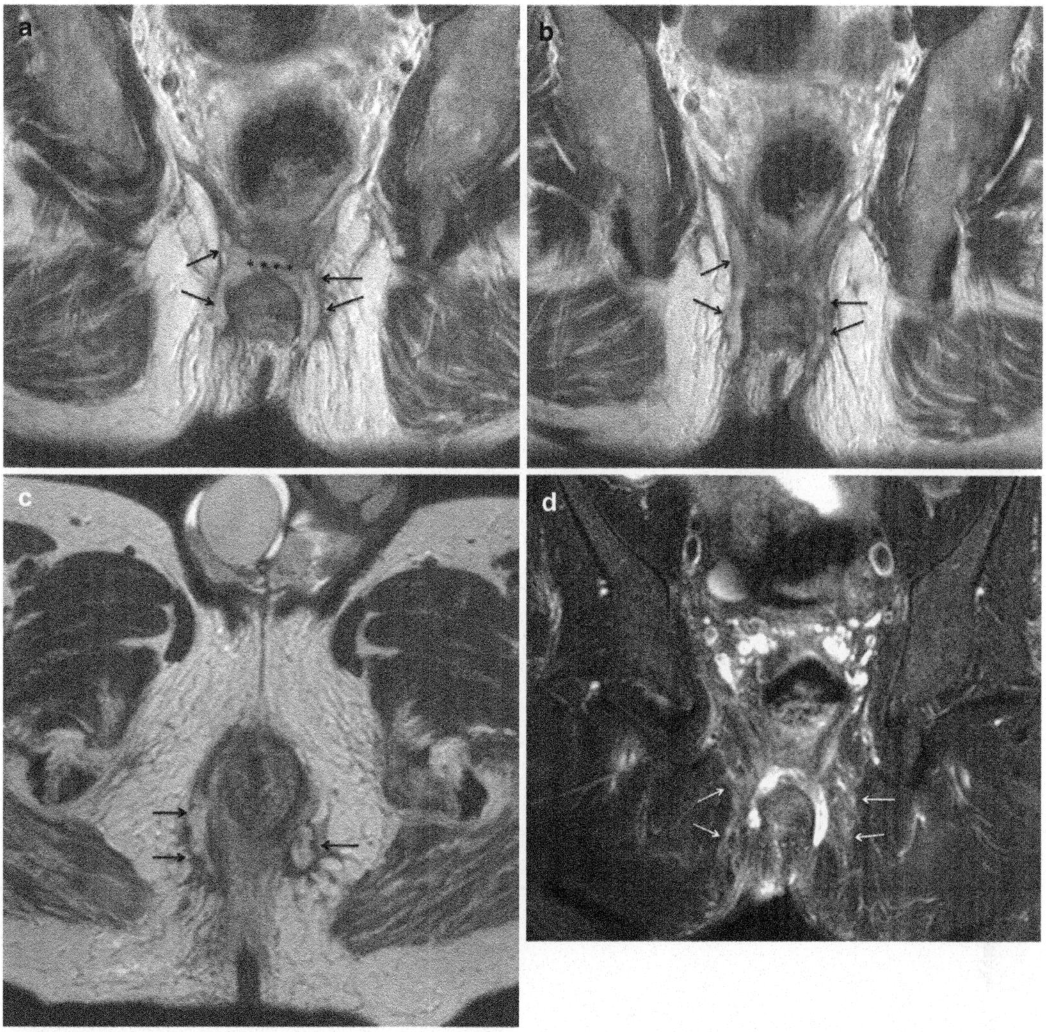

Fig. 9.8 Complex perianal fistula. Coronal-oblique T2-weighted TSE images (**a**, **b**) and axial-oblique T2-weighted TSE image (**c**) show a horseshoe-like intersphincteric fistulous tract (asterisks in **a**), which extent bilaterally toward ischioanal fossa through external sphincter (arrows). Coronal-oblique T2-weighted TSE fat-saturated image (**d**) well demonstrates perifistulous inflammatory edema (arrows)

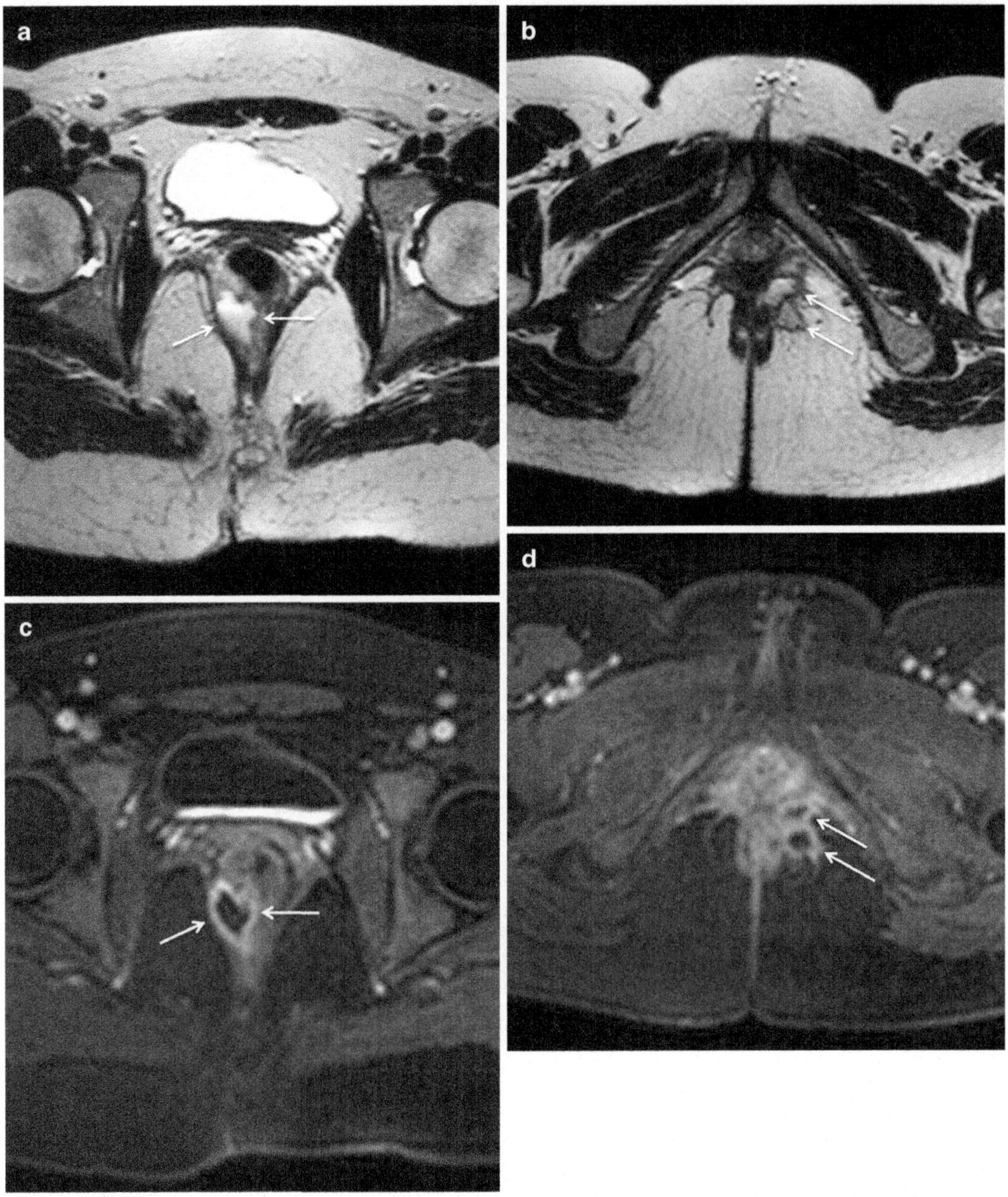

Fig. 9.9 Perianal fistula with multiple abscesses. Axial-oblique T2-weighted TSE image (**a**) shows a right-sided abscess (arrows). A more caudal scan (**b**) demonstrates other two small abscesses in the left ischioanal fossa (arrows). Axial-oblique T1-weighted TSE fat-saturated images, obtained after i.v. injection of Gadolinium (**c**, **d**) shows peripheral contrast enhancement of abscesses, containing a central component of non-enhancing purulent material (arrows)

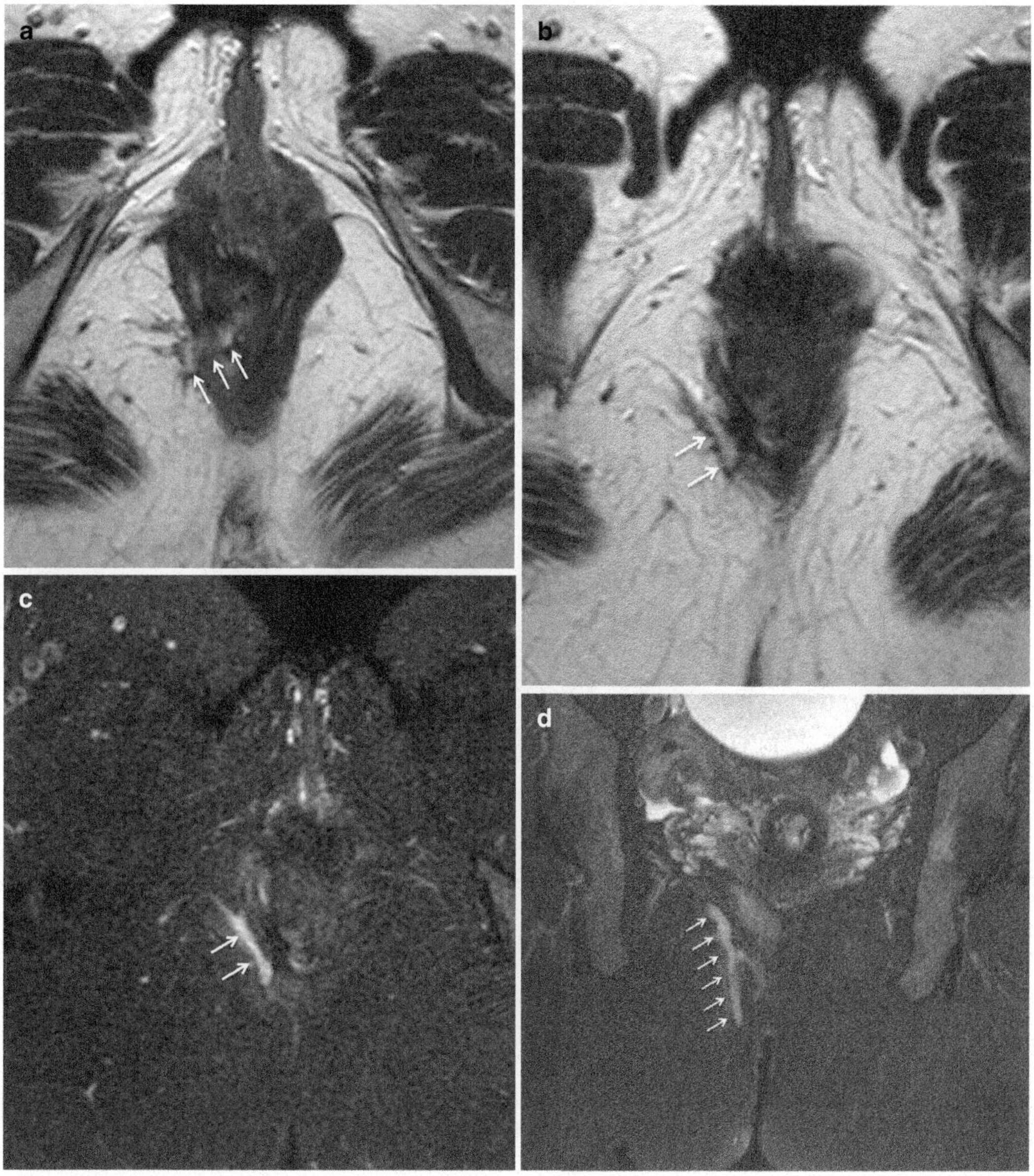

Fig. 9.10 Trans-sphincteric fistula. Axial-oblique T2-weighted TSE image (**a**), showing the internal opening of fistula at 7 o'clock (arrows). More caudally, T2-weighted TSE images, performed without (**b**) and with fat saturation (**c**), show a fistulous tract in the right ischioanal fossa (arrows). Coronal-oblique T2-weighted TSE fat-saturated image (**d**) clearly traces the craniocaudal extent of fistula (arrows)

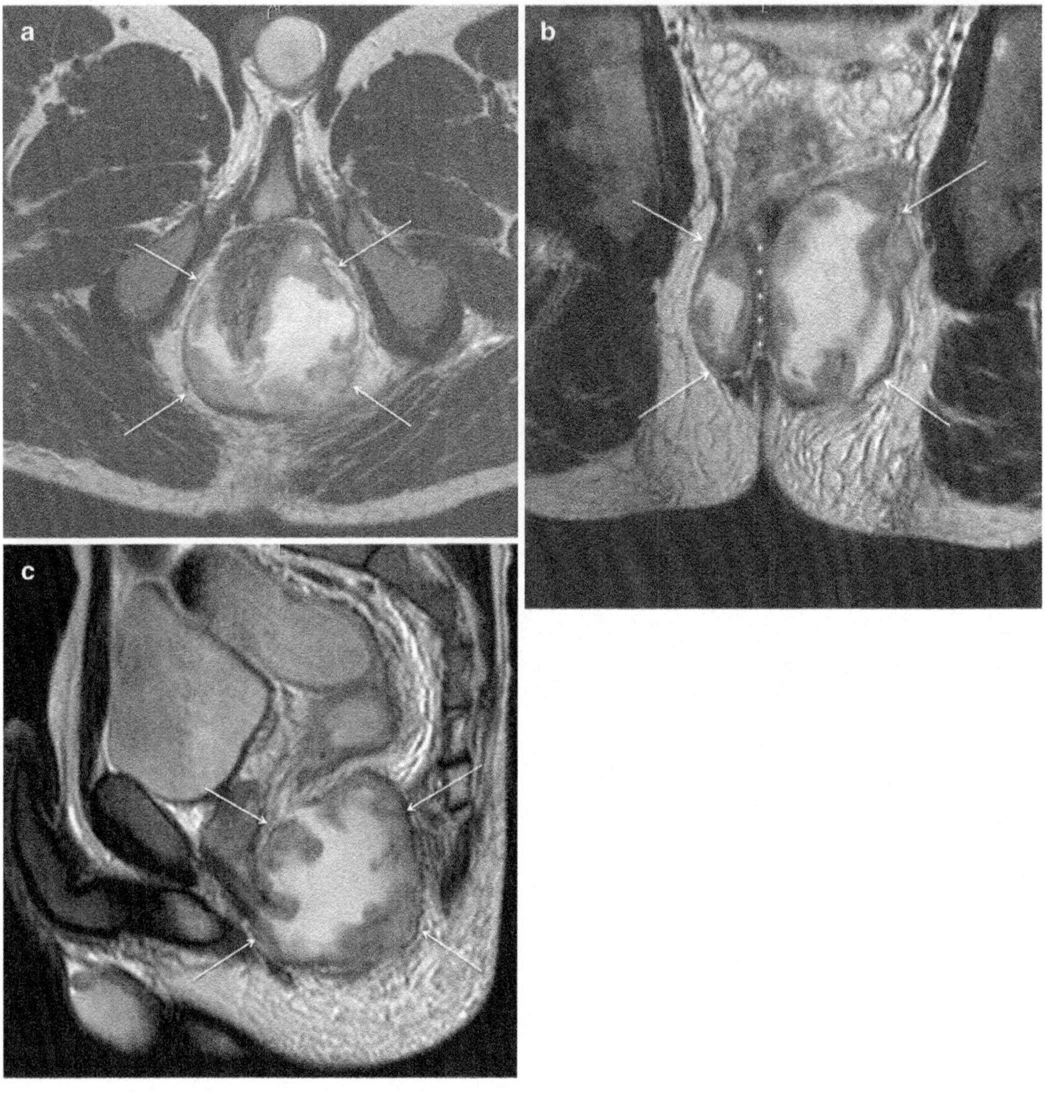

Fig. 9.11 Perianal abscess. Axial-oblique (**a**), coronal-oblique (**b**), and sagittal (**c**) T2-weighted TSE images show a voluminous perianal abscess (arrows) that displaces to the right side of the sphincteric complex (asterisks)

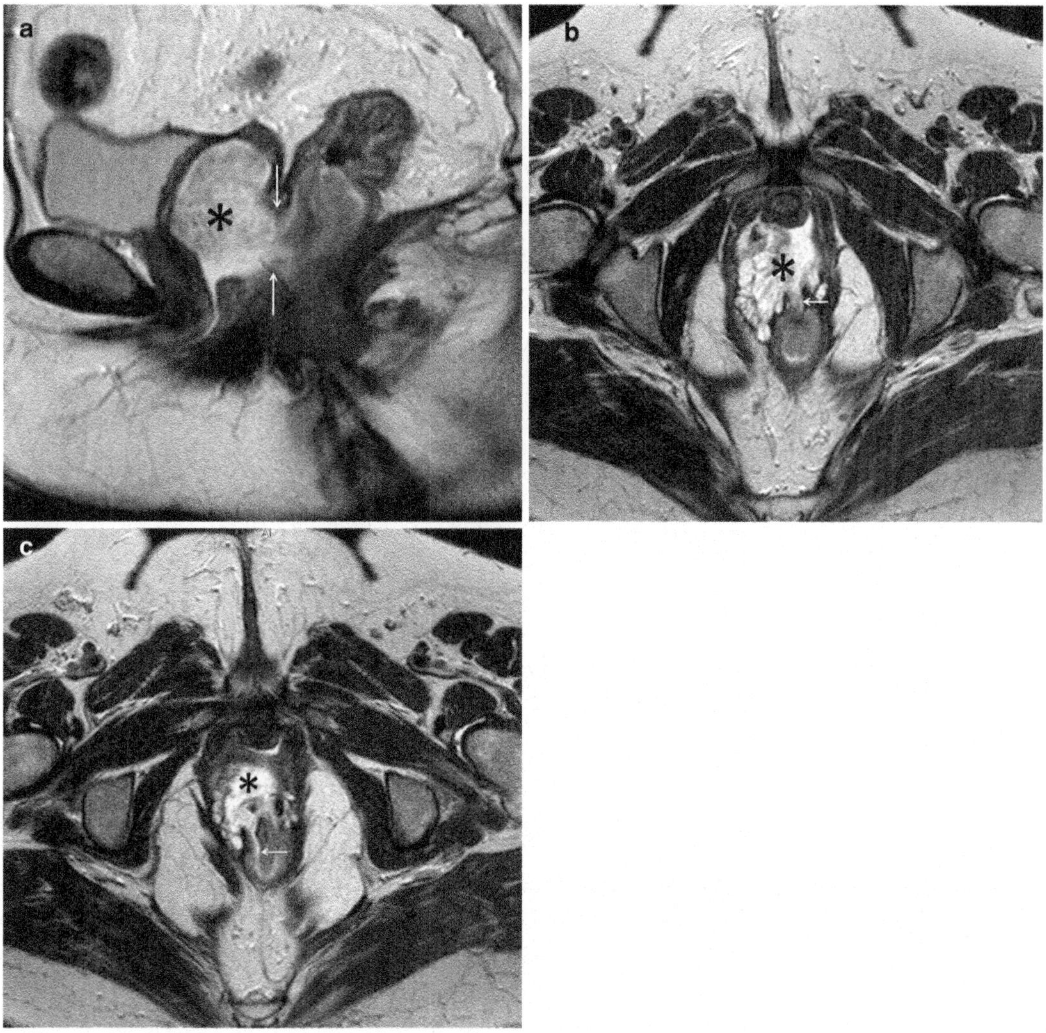

Fig. 9.12 Recto-vaginal communication. Hysterectomized patient. Sagittal (**a**) and axial-oblique (**b**, **c**) T2-weighted TSE images clearly depict a wide defect of the posterior vaginal wall (arrows in a). On axial-oblique images (**b**, **c**) some fistulous tracts demonstrate a communication with the anterior wall of rectum (arrows). Vaginal lumen is filled with fecaloid material (asterisk)

9.4 MRI of the Anal Canal beyond Crohn's Disease

A number of diseases can take place within the perianal region.

Beyond congenital malformations, they can be grouped into three main types: inflammatory conditions, benign, and malignant lesions.

MRI protocol is the same used for evaluation of fistula-in-ano generally performed in CD patients.

The main purpose of the exam is to distinguish CD involvement of the anal canal and perianal structures from a different pathology.

9.4.1 Inflammatory Conditions

9.4.1.1 Pilonidal Sinus

Pilonidal sinus derives from a folliculitis at the intergluteal natal cleft and leads to the formation of a sinus tract or abscess [20, 25–27].

It is typically located within the subcutaneous soft tissues near the sacrum or the coccyx and it is characterized by hyperintensity on T2-weighted images [28].

Differential diagnosis includes fistula-in-ano, unlike which no involvement of the sphincteric space nor communication with the anal lumen can be detected [20, 25–28].

9.4.1.2 Hidradenitis Suppurativa

Hidradenitis suppurativa is caused by a chronic occlusion of cutaneous hair follicles of apocrine sweat glands which lead to recurrent inflammations, sinus tract and abscesses formation, and fibrosis [20, 26, 27].

The perineum represents the typical localization in males, whereas the axillae are in females [25].

Inflamed structures appear hyperintense on T2-weighted sequences, particularly on the fat-saturated ones, and surrounded by edema of the adjacent fat tissue [20, 26].

Contrast medium injection may be helpful in demonstrating rim enhancement of abscessual collections [25].

Differentiation from perianal involvement in CD patients may be challenging, since the two conditions may coexist [25].

Extensive bilateral involvement, absence of rectal wall thickening, and lack of communication with the anal canal are findings suggestive of hidradenitis suppurativa [25, 26].

9.4.1.3 Fournier Gangrene

Fournier gangrene is a urological emergency consisting of necrotizing fasciitis of the genitourinary, perineal, and perianal tissues caused by infection of both aerobic and anaerobic bacteria.

Common causes are traumas or perianal fistulas, although etiology may remain unclear in some cases. Risk factors include diabetes mellitus, obesity, and immunodeficiency.

This condition is burdened by high mortality rates. Death may occur mainly due to sepsis and multiple organ failure [29, 30].

MRI takes advantage of high soft-tissue contrast resolution and a wide field of view.

Moreover, the performance of DWI scans allows recognition of high cellular-density processes, such as abscesses, without any need for intravenous contrast medium injection.

Hence, MRI can be successfully applied not only for initial diagnosis but also during follow-up in order to evaluate treatment outcomes, although its role is usually limited due to prolonged scan times and usual lower scanner availability.

9.4.2 Benign Lesions

Benign lesions of the perianal region can be grossly divided into cystic and fat containing.

Epidermoid and dermoid cysts arise from the ectoderm. While the formers usually appear as unilocular fluid-filled lesions, the latter are more heterogeneous at MRI due to entrapment of ectodermal elements.

As dermoid cysts, also enteric developmental cysts are characterized by inhomogeneous signal intensity. They represent rare congenital malformations of the developing digestive tract and are of endodermal origin [28].

Fat-containing lesions include lipomas and teratomas.

Lipomas are characterized by homogeneous hyperintensity on T1-weighted images with drop of signal on fat-saturated scans [28].

On the other hand, the adipose content within presacral teratoma is lower since fluid and solid components usually coexist. Moreover, intralesional septations and calcifications are not rare [28, 31].

9.4.3 Malignant Neoplasms

Anal canal carcinomas represent only a small percentage (2.5%) of the whole colorectal carcinomas [25, 26, 32].

The main histological type is squamous cell carcinoma (SSC; 80–85%), whereas the remaining include adenocarcinomas, melanomas, lymphoma, GI stromal tumor, and leiomyosarcomas [25, 26, 32].

Risk factors for SCC occurrence are chronic infection with human papillomavirus, in particular 16 and 18 types, human immunodeficiency virus (HIV), infections, and immunosuppression [25, 32].

Resort to MRI is necessary for local staging, including tumor size, sphincteric infiltration, and lymph nodal spread.

At MRI, SCC shows itself as a lobulated intraluminal or infiltrative mass characterized by iso/hypointense on T1-weighted images, iso/hyperintense on the T2-weighted ones, enhancement after intravenous contrast agent injection, and water restriction on DWI scans [25, 26, 32].

References

1. Schwartz DA, Pemberton JH, Sandborn WJ. Diagnosis and treatment of perianal fistulas in Crohn disease. Ann Intern Med. 2001;135:906–18.
2. Makowiec F, Jehle EC, Becker HD, et al. Perianal abscess in Crohn's disease. Dis Colon Rectum. 1997;40:443–50.
3. Rankin GB, Watts HD, Melnyk CS, et al. National cooperative Crohn's disease study: extraintestinal manifestations and perianal complications. Gastroenterology. 1979;77:914–20.
4. Williams DR, Coller JA, Corman ML, et al. Anal complications in Crohn's disease. Dis Colon Rectum. 1981;24:22–4.
5. Buchmann P, Keighley MR, Thompson H, et al. Natural history of perianal Crohn's disease. Ten year follow-up: a plea for conservatism. Am J Surg. 1980;140:642–4.
6. van Dongen LM, Lubbers EJ. Perianal fistulas in patients with Crohn's disease. Arch Surg. 1986;121:1187–890.
7. Goebell H. Perianal complications in Crohn's disease. Neth J Med. 1990;37(Suppl 1):S47–51.
8. Fielding JF. Perianal lesions in Crohn's disease. J R Coll Surg Edinb. 1972;17:32–7.
9. American Gastroenterological Association. AGA technical review on perianal Crohn's disease. Gastroenterology. 2003;125:1508–30.
10. Hellers G, Bergstrand O, Ewerth S, et al. Occurrence and outcome after primary treatment of anal fistulae in Crohn's disease. Gut. 1980;21:525–7.
11. Safar B, Sands D. Perianal Crohn's disease. Clin Colon Rectal Surg. 2007;20:282–93.
12. Parks A. The pathogenesis and treatment of fistula-in-ano. Br Med J. 1961;1:463–9.
13. Lockhart-Mummery HE. Symposium. Crohn's disease: anal lesions. Dis Colon Rectum. 1975;18:200–2.
14. Alabaz O, Weiss EG. Anorectal Crohn's disease. In: Beck D, Wexner SD, editors. Fundamentals of anorectal surgery. 2nd ed. Philadelphia: WB Saunders; 1999. p. 498–509.
15. Maccioni F, Colaiacomo MC, Stasolla A, et al. Value of MRI performer with phased-array coil in the diagnosis and pre-operative classification of perianal and anal fistula. Radiol Med. 2002;104:58.
16. Parks AG, Gordon PH, et al. A classification of fistula-in-ano. Br J Surg. 1976;63:1–12.
17. Morris J, Spencer JA, Ambrose NS. MR imaging classification of perianal fistulas and its implications for patient management. Radiographics. 2000;20(3):623–35.; discussion 635-7 PMID: 10835116. https://doi.org/10.1148/radiographics.20.3.g00mc15623.
18. Whiteford MH, Kilkenny J 3rd, Hyman N, Buie WD, Cohen J, Orsay C, Dunn G, Perry WB, Ellis CN, Rakinic J, Gregorcyk S, Shellito P, Nelson R, Tjandra JJ, Newstead G, Standards Practice Task Force, American Society of Colon and Rectal Surgeons. Practice parameters for the treatment of perianal abscess and fistula-in-ano (revised). Dis Colon Rectum. 2005;48(7):1337–42. https://doi.org/10.1007/s10350-005-0055-3. PMID: 15933794
19. Sheedy SP, Bruining DH, Dozois EJ, Faubion WA, Fletcher JG. MR imaging of perianal Crohn disease. Radiology. 2017;282(3):628–45. https://doi.org/10.1148/radiol.2016151491. PMID: 28218881
20. Cicero G, Ascenti G, Blandino A, Pallio S, Abate C, D'Angelo T, Mazziotti S. Magnetic resonance imaging of the anal region: clinical applications. J Clin Imaging Sci. 2020;10:76. https://doi.org/10.25259/JCIS_180_2020. PMID: 33274120; PMCID: PMC7708963

21. de Miguel CJ, del Salto LG, Rivas PF, del Hoyo LF, Velasco LG, de las Vacas MI, Marco Sanz AG, Paradela MM, Moreno EF. MR imaging evaluation of perianal fistulas: spectrum of imaging features. Radiographics. 2012;32(1):175–94. https://doi.org/10.1148/rg.321115040. PMID: 22236900
22. O'Malley RB, Al-Hawary MM, Kaza RK, Wasnik AP, Liu PS, Hussain HK. Rectal imaging: part 2, Perianal fistula evaluation on pelvic MRI—what the radiologist needs to know. AJR Am J Roentgenol. 2012;199(1):W43–53. https://doi.org/10.2214/AJR.11.8361. PMID: 22733931
23. Halligan S, Stoker J. Imaging of fistula in ano. Radiology. 2006;239(1):18–33. https://doi.org/10.1148/radiol.2391041043. PMID: 16567481
24. Cicero G, Mazziotti S. Crohn's disease at radiological imaging: focus on techniques and intestinal tract. Intest Res. 2021;19(4):365–78. https://doi.org/10.5217/ir.2020.00097. Epub 2020 Nov 25. PMID: 33232590; PMCID: PMC8566824
25. Erlichman DB, Kanmaniraja D, Kobi M, Chernyak V. MRI anatomy and pathology of the anal canal. J Magn Reson Imaging. 2019;50(4):1018–32. https://doi.org/10.1002/jmri.26776. Epub 2019 May 22. PMID: 31115134
26. Balcı S, Onur MR, Karaosmanoğlu AD, Karçaaltıncaba M, Akata D, Konan A, Özmen MN. MRI evaluation of anal and perianal diseases. Diagn Interv Radiol. 2019;25(1):21–7. https://doi.org/10.5152/dir.2018.17499. PMID: 30582572; PMCID: PMC6339630
27. Erden A. MRI of anal canal: normal anatomy, imaging protocol, and perianal fistulas: part 1. Abdom Radiol (NY). 2018;43(6):1334–52. https://doi.org/10.1007/s00261-017-1305-2. PMID: 28840368
28. O'Neill DC, Murray TE, Thornton E, Burke J, Dunne R, Lee MJ, Morrin MM. Imaging features of benign perianal lesions. J Med Imaging Radiat Oncol. 2019;63(5):617–23. https://doi.org/10.1111/1754-9485.12934. Epub 2019 Aug 1. PMID: 31368659
29. El-Qushayri AE, Khalaf KM, Dahy A, Mahmoud AR, Benmelouka AY, Ghozy S, Mahmoud MU, Bin-Jumah M, Alkahtani S, Abdel-Daim MM. Fournier's gangrene mortality: a 17-year systematic review and meta-analysis. Int J Infect Dis. 2020;92:218–25. https://doi.org/10.1016/j.ijid.2019.12.030. Epub 2020 Jan 18. PMID: 31962181
30. Levenson RB, Singh AK, Novelline RA. Fournier gangrene: role of imaging. Radiographics. 2008;28(2):519–28. https://doi.org/10.1148/rg.282075048. PMID: 18349455
31. Tappouni RF, Sarwani NI, Tice JG, Chamarthi S. Imaging of unusual perineal masses. AJR Am J Roentgenol. 2011;196(4):W412–20. https://doi.org/10.2214/AJR.10.4728. PMID: 21427305
32. Erden A. MRI of anal canal: common anal and perianal disorders beyond fistulas: part 2. Abdom Radiol (NY). 2018;43(6):1353–67. https://doi.org/10.1007/s00261-017-1306-1.

Structured Reporting

10

Tommaso D'Angelo, Thomas J. Vogl, Simon S. Martin, Ibrahim Yel, and Christian Booz

MR-Enterography and perianal MRI reports must guide clinicians and surgeons in establishing the proper treatment.

However, owing to the different clinical scenarios and the number of CD features, including all the necessary information may seem a challenging task.

In this context, structured reports represent a practical choice for both radiologists and physicians, in terms of clarity of information and avoidance of superfluous data [1–3].

Currently, several templates have been proposed by different research groups [4–6].

T. D'Angelo (✉)
Department of Biomedical Sciences and Morphological and Functional Imaging, University Hospital Messina, Messina, Italy
e-mail: tdangelo@unime.it

T. J. Vogl
Department of Diagnostic and Interventional Radiology, University Hospital Frankfurt, Frankfurt am Main, Germany
e-mail: t.vogl@em.uni-frankfurt.de

S. S. Martin · I. Yel
Department of Diagnostic and Interventional Radiology, University Hospital Frankfurt, Frankfurt am Main, Germany

Division of Experimental Imaging, Department of Diagnostic and Interventional Radiology, University Hospital Frankfurt, Frankfurt am Main, Germany
e-mail: simon.martin@kgu.de; Ibrahim.Yel@kgu.de

C. Booz
Division of Experimental Imaging, Department of Diagnostic and Interventional Radiology, University Hospital Frankfurt, Frankfurt am Main, Germany
e-mail: christian.booz@kgu.de

10.1 MR-Enterography

MRE is a safe and minimally invasive modality for small bowel assessment.

Therefore, MRE report must represent a summary of the radiologic findings essential for the clinician in order to establish the subsequent patient's management.

First, every report should include the amount of enteric contrast agent administrated to the patient and information about its progression and intestinal motility through the small and large bowel loops.

When present, luminal narrowing of the small bowel have to be recognized and quantified as well as the upstream loops dilation (mild: 3–4 cm; moderate/severe: >4 cm) [7].

In CD patients, wall thickening (or thickenings, if multiple skip lesions are found) must be identified, anatomically localized according to the intestinal segment and the abdominal quadrant involved, and measured through digital calipers in length and thickness (mild 3–5 mm; moderate: 5–9 mm; severe: ≥10 mm) [5, 7].

Information from the different sequences of MRE protocol have to be combined in order to specify the disease pattern (active inflammatory, fistulizing/perforating, fibrostenotic, reparative/regenerative) [8].

Intestinal lesions have to be assessed on the basis of signal intensity on T2-weighted scans (with and without fat saturation), in order to detect mural edema or intramural fat deposition.

S. Mazziotti et al. (eds.), *MR Enterography*, https://doi.org/10.1007/978-3-031-11930-9_10

Presence of inflammation is revealed by hyperintensity on DWI scans and drop of signal on the relative ADC map. So far, no established thresholds of ADC values have been found for differentiating active inflammation from fibrosis, therefore the evaluation of these acquisitions remains qualitative [7].

In case of penetrating disease pattern, irregularity of the external wall layer may indicate the presence of sinus tracts or fistulas, identifiable as tubular structures developing within the mesenteric fat or connecting the enteric lumen with an extraintestinal structure [7].

Comparison with the previous examination is of the utmost importance in demonstrating therapy response.

Extra-luminal inflammatory-related findings, such as perivisceral edema, comb sign, creeping fat, enlarged lymph nodes, and free abdominal fluid, must be acknowledged [7].

Complications can be ascribed to primary disease or treatment and can be appreciated within the mesenteric adipose tissue (abscesses), abdominal organs (primary sclerosing cholangitis, pancreatitis, renal lithiasis), skeletal structures (sacroiliitis), or cutaneous surface.

Table 10.1 Example of MRE report for CD patients

- **Amount of enteric contrast medium administrated (.....ml)**
- **Progression of enteric contrast medium:**
  - Regular/irregular
  - Fast
  - Delayed
- **Bowel wall thickening(s):**
  - Number and localization (intestinal segment; abdominal quadrant)
  - Maximum length (…mm)
  - Thickness (mild 3–5 mm; moderate: 5–9 mm; severe: >10 mm)
- **Intestinal signs of active inflammation:**
  - Mural edema
  - Contrast hyperenhancement (homogeneous, asymmetric, layered)
  - Water restriction on DWI scans
- **Mesenteric signs of inflammation:**
  - Comb sign
  - Enlarged lymph nodes (short axis maximum diameter)
  - Free abdominal fluid
  - Fibrofatty proliferation
- **Luminal narrowing:**
  - Stricture localization (intestinal segment; abdominal quadrant)
  - Upstream dilation (mild: 3–4 cm; moderate/severe: >4 cm)
- **Penetrating disease:**
  - Sinus tract (site, extension)
  - Fistula (site, extension, structures connected)
  - Abscess (site, dimensions)
- **Extra-intestinal findings**
  - Abdominal organs
  - Musculoskeletal structures
  - Cutaneous
  - Other
- **Comparison with previous exams:**
  - Response to therapy (yes/no)

In patients who have already undergone surgery, the type of operation, localization, and patency of intestinal anastomosis or cutaneous stomia must be described.

In Table 10.1, it is provided a structured report generally used in our department.

10.2 Perianal MRI

The main goal of a radiologic report concerning the anal canal and the perianal region is to provide an overview of the pathologic findings in order to establish the patient's following conservative or invasive treatment.

In the latter case, the report must serve as a guide for the surgical approach.

First, considering that the patient's decubitus may vary among the clinical examination (prone), the intraoperative assessment (lithotomy), and the MR scanning position (usually supine), the latter must be clarified in the report [9].

Fistulas are generally recognized as tubular or serpiginous tracts filled with fluid or pus and are characterized by hyperintensity on T2-weighted images, especially on the fat-saturated ones [9, 10].

The coexistence of T2-hyperintensity and peripheral rim enhancement within a fistula reveals the presence of granulation tissue as a sign of healing process [10, 11].

On the contrary, features of fibrotic involution include T2-hypointensity and absence or persistence of progressive contrast enhancement [10].

Beyond the established role of DWI scans in identifying fistulas, this type of sequence seems to have a role in differentiating active from inactive tracts owing to the different ADC values (lower in case of activity) [12, 13].

Pathological findings of the anal region have to be described according to a three-part division of the anal canal (upper, middle, and lower third) and to a clock face scheme [9].

The "anal clock" system is applied to axial images and it is used to describe the origin and the route of the fistula; at supine scanning position the perineum is located at 12 o'clock, while the intergluteal cleft is at 6 o'clock, the right and the left ischioanal fossae are sited, respectively, at 9 and 3 o'clock [10, 14–18].

When a fistula is present, the first finding to describe is the internal opening, namely the communication between the fistula and the anal lumen.

Although measurements can be made using calipers on coronal-oblique images (i.e., from the internal opening to the anal verge), they are not mandatory, since distances can vary on the basis of the patient's position during scanning and surgical operation [9, 10].

Afterward, the path of the fistula should be reported on the basis of Parks or St. James's University Hospital classifications [14].

The relationship between the track and the sphincteric and/or muscular structures as well as the adipose tissue of ischioanal and gluteal regions must be reported [14].

Fistulas may also present single or multiple secondary tracts (or branches) or communication with an abscessual collection [10, 11].

The localization of the distal opening is extremely variable: if cutaneous, most of the time it is placed near the anal verge; otherwise, if the route of the fistula is anterior, it can develop through the posterior wall of the vagina and communicate with the vaginal lumen.

On the other hand, if the tract is blind-ended, it takes the name of "sinus tract" [10].

T2-hyperintensity can be appreciated within the perilesional fat tissue, indicating edema [11].

When the inflammatory process is extensive, it can spread to the rectum with thickening and edema of the walls (proctitis) [10, 15].

In patients who already undergone invasive treatment, surgical drainages, fat packing, and fibrotic stripes can also be visualized. The pres-

Table 10.2 Example of a reporting template of perianal fistulas

Main track:
• Internal opening localization (anal clock; anal canal third)
• Maximum axial caliber (…mm).
• Route:
– Inter-sphincteric
– Trans-sphincteric
– Extra-sphincteric
– Supra-sphincteric
– Superficial
• Anatomic relationships with:
– Sphincterial structures
– Muscular structures
– Adipose tissue (ischioanal/gluteal)
• Distal end:
– Blind-ended (sinus tract)
– Cutaneous opening (fistula)
– Localization
• Radiological signs of activity
– T2-w hyperintensity
– Contrast hyperenhancement
– Water restriction on DWI
• Radiological signs of fibrosis
– T2-w hypointensity
– Absent/progressive contrast enhancement
Secondary branches:
• Number
• Localization (anal clock)
• Path
• Distal end
• Signs of activity/fibrosis
Abscess:
• Number
• Localization
• Dimensions
• Communication with a fistula
Anovaginal fistula:
• Anal origin (anal clock; anal canal third)
• Path
• Distal end (vaginal portion involved)
Extra-anal extension:
• Levator ani injury
• Signs of proctitis (edema; wall thickening)

ence of setons can be appreciated as focal filling defects within a fistula [11].

In Table 10.2, a structured report for perianal involvement in CD is summarized.

References

1. Gomez AA, Nunes TF, dos Santos CHM, Rissato DM, Tibanaa TK, Adôrno IF, Dourado DM. Comparison of conventional and structured report in the evaluation of Crohn's disease through enterography. J Coloproctol (rio j). 2018;38(4):290–4.
2. Wildman-Tobriner B, Allen BC, Bashir MR, Camp M, Miller C, Fiorillo LE, Cubre A, Javadi S, Bibbey AD, Ehieli WL, McGreal N, Quevedo R, Thacker JK, Mazurowski M, Jaffe TA. Structured reporting of CT enterography for inflammatory bowel disease: effect on key feature reporting, accuracy across training levels, and subjective assessment of disease by referring physicians. Abdom Radiol (NY). 2017;42(9):2243–50. https://doi.org/10.1007/s00261-017-1136-1. PMID: 28393301.
3. Wildman-Tobriner B, Allen BC, Davis JT, Miller CM, Schooler GR, McGreal NM, Quevedo R, Thacker JK, Jaffe TA. Structured reporting of magnetic resonance enterography for pediatric Crohn's disease: effect on key feature reporting and subjective assessment of disease by referring physicians. Curr Probl Diagn Radiol. 2017;46(2):110–4. https://doi.org/10.1067/j.cpradiol.2016.12.001. Epub 2016 Dec 5
4. Baker ME, Fletcher JG, Al-Hawary M, Bruining D. Interdisciplinary updates in Crohn's disease reporting nomenclature, and cross-sectional disease monitoring. Radiol Clin N Am. 2018;56(5):691–707. https://doi.org/10.1016/j.rcl.2018.04.010.
5. Bruining DH, Zimmermann EM, Loftus EV Jr, Sandborn WJ, Sauer CG, Strong SA, Society of Abdominal Radiology Crohn's Disease-Focused Panel. Consensus recommendations for evaluation, interpretation, and utilization of computed tomography and magnetic resonance Enterography in patients with small bowel Crohn's disease. Radiology. 2018;286(3):776–99. https://doi.org/10.1148/radiol.2018171737. Epub 2018 Jan 10
6. Stanzione A, Boccadifuoco F, Cuocolo R, Romeo V, Mainenti PP, Brunetti A, Maurea S. State of the art in abdominal MRI structured reporting: a review. Abdom Radiol (NY). 2021;46(3):1218–28. https://doi.org/10.1007/s00261-020-02744-8. Epub 2020 Sep 16. PMID: 32936418; PMCID: PMC7940284
7. Guglielmo FF, Anupindi SA, Fletcher JG, Al-Hawary MM, Dillman JR, Grand DJ, Bruining DH, Chatterji M, Darge K, Fidler JL, Gandhi NS, Gee MS, Grajo JR, Huang C, Jaffe TA, Park SH, Rimola J, Soto JA, Taouli B, Taylor SA, Baker ME. Small bowel Crohn disease at CT and MR enterography: imaging atlas and glossary of terms. Radiographics. 2020;40(2):354–75. https://doi.org/10.1148/rg.2020190091. Epub 2020 Jan 17
8. Kaushal P, Somwaru AS, Charabaty A, Levy AD. MR enterography of inflammatory bowel disease with endoscopic correlation. Radiographics. 2017;37(1):116–31. https://doi.org/10.1148/rg.2017160064. Epub 2016 Nov 25

9. Ho E, Rickard MJFX, Suen M, Keshava A, Kwik C, Ong YY, Yang J. Perianal sepsis: surgical perspective and practical MRI reporting for radiologists. Abdom Radiol (NY). 2019;44(5):1744–55. https://doi.org/10.1007/s00261-019-01920-9.
10. Thipphavong S, Costa AF, Ali HA, Wang DC, Brar MS, Jhaveri KS. Structured reporting of MRI for perianal fistula. Abdom Radiol (NY). 2019;44(4):1295–305. https://doi.org/10.1007/s00261-018-1839-y.
11. O'Malley RB, Al-Hawary MM, Kaza RK, Wasnik AP, Liu PS, Hussain HK. Rectal imaging: part 2, Perianal fistula evaluation on pelvic MRI—what the radiologist needs to know. AJR Am J Roentgenol. 2012;199(1):W43-53. https://doi.org/10.2214/AJR.11.8361.
12. Boruah DK, Hazarika K, Ahmed H, Borah KK, Borah S, Malakar S, Hajoari N. Role of diffusion-weighted imaging in the evaluation of perianal fistulae. Indian J Radiol Imaging. 2021;31(1):91–101. https://doi.org/10.1055/s-0041-1729673. Epub 2021 May 23. PMID: 34316116; PMCID: PMC8299510
13. Yoshizako T, Wada A, Takahara T, Kwee TC, Nakamura M, Uchida K, Hara S, Luijten PR, Kitagaki H. Diffusion-weighted MRI for evaluating perianal fistula activity: feasibility study. Eur J Radiol. 2012;81(9):2049–53. https://doi.org/10.1016/j.ejrad.2011.06.052. Epub 2011 Jul 20. PMID: 21767926
14. de Miguel CJ, del Salto LG, Rivas PF, del Hoyo LF, Velasco LG, de las Vacas MI, Marco Sanz AG, Paradela MM, Moreno EF. MR imaging evaluation of perianal fistulas: spectrum of imaging features. Radiographics. 2012;32(1):175–94. https://doi.org/10.1148/rg.321115040.
15. Cicero G, Ascenti G, Blandino A, Pallio S, Abate C, D'Angelo T, Mazziotti S. Magnetic resonance imaging of the anal region: clinical applications. J Clin Imaging Sci. 2020;10:76. https://doi.org/10.25259/JCIS_180_2020. PMID: 33274120; PMCID: PMC7708963
16. Santillan CS, Huang C, Eisenstein S, Al-Hawary MM. MRI of perianal Crohn disease: technique and interpretation. Top Magn Reson Imaging. 2021;30(1):63–76. https://doi.org/10.1097/RMR.0000000000000268.
17. Mazziotti S, Ascenti G, Scribano E, Gaeta M, Pandolfo A, Bombaci F, Donato R, Fries W, Blandino A. Guide to magnetic resonance in Crohn's disease: from common findings to the more rare complicances. Inflamm Bowel Dis. 2011;17(5):1209–22. https://doi.org/10.1002/ibd.21548. Epub 2010 Nov 5
18. Cicero G, Mazziotti S. Crohn's disease at radiological imaging: focus on techniques and intestinal tract. Intest Res. 2021;19(4):365–78. https://doi.org/10.5217/ir.2020.00097. Epub 2020 Nov 25. PMID: 33232590; PMCID: PMC8566824

11 Clinical Cases

Giuseppe Cicero, Tommaso D'Angelo,
Alfredo Blandino, and Silvio Mazziotti

11.1 MR-Enterography Clinical Cases (Figs. 11.1, 11.2, 11.3, 11.4, 11.5, 11.6, 11.7, 11.8, 11.9, 11.10, 11.11, 11.12, 11.13, 11.14, 11.15, 11.16, 11.17, 11.18, 11.19, 11.20, 11.21, 11.22 and 11.23)

G. Cicero (✉)· T. D'Angelo · A. Blandino
S. Mazziotti
Department of Biomedical Sciences and
Morphological and Functional Imaging, University
Hospital Messina, Messina, Italy
e-mail: gcicero@unime.it; tdangelo@unime.it;
ablandino@unime.it; smazziotti@unime.it

S. Mazziotti et al. (eds.), *MR Enterography*, https://doi.org/10.1007/978-3-031-11930-9_11

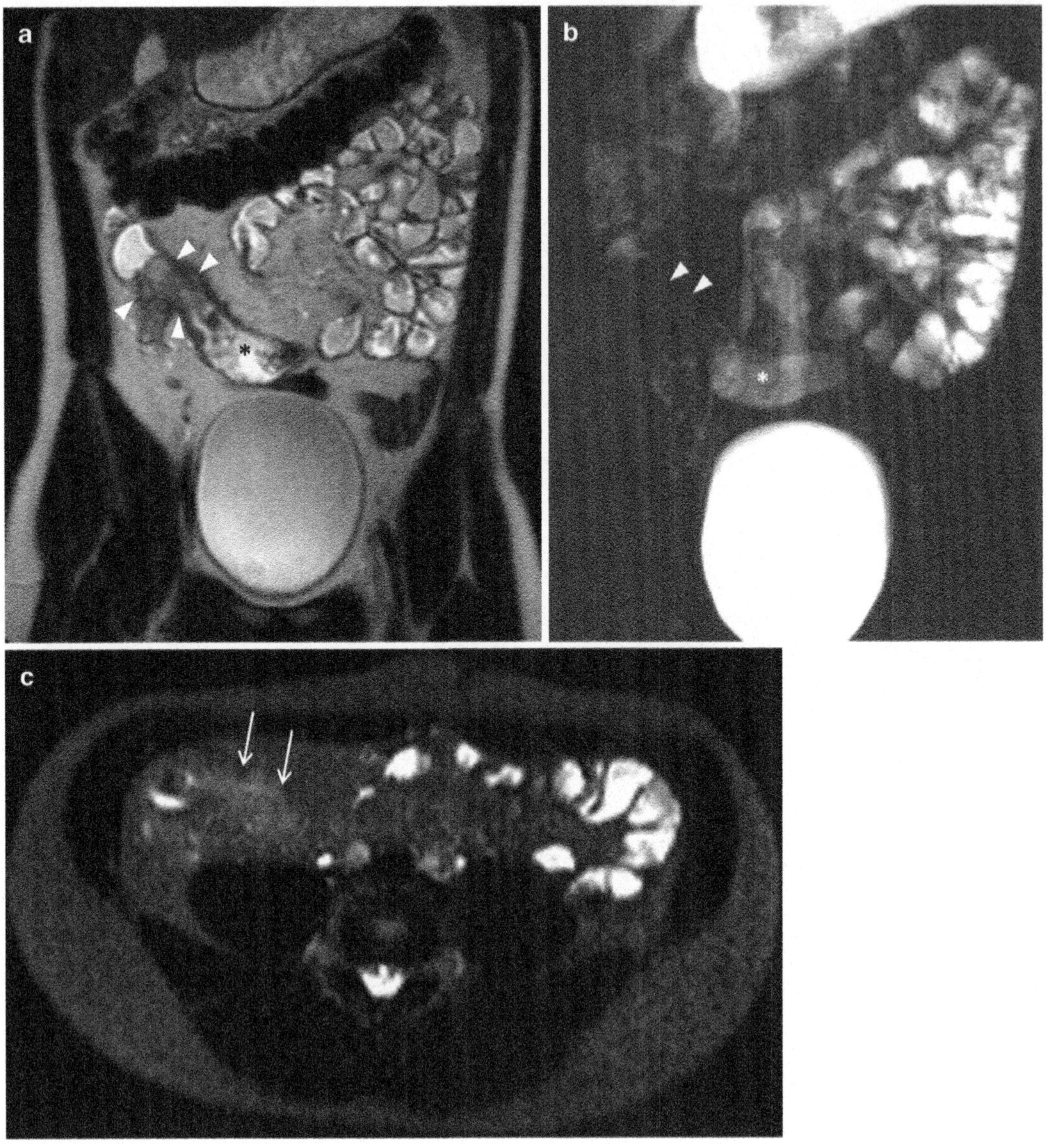

Fig. 11.1 A 17-year-old male patient suffering from CD. Coronal T2-w HASTE (**a**) and coronal T2-w RARE thick-slab (**b**) images show a severe thickening of the terminal ileum extending into the cecum with luminal narrowing (arrowheads) and dilation of the upstream loop (asterisk). Mural edema (arrows) is appreciable on axial T2-w HASTE with fat saturation (**c**)

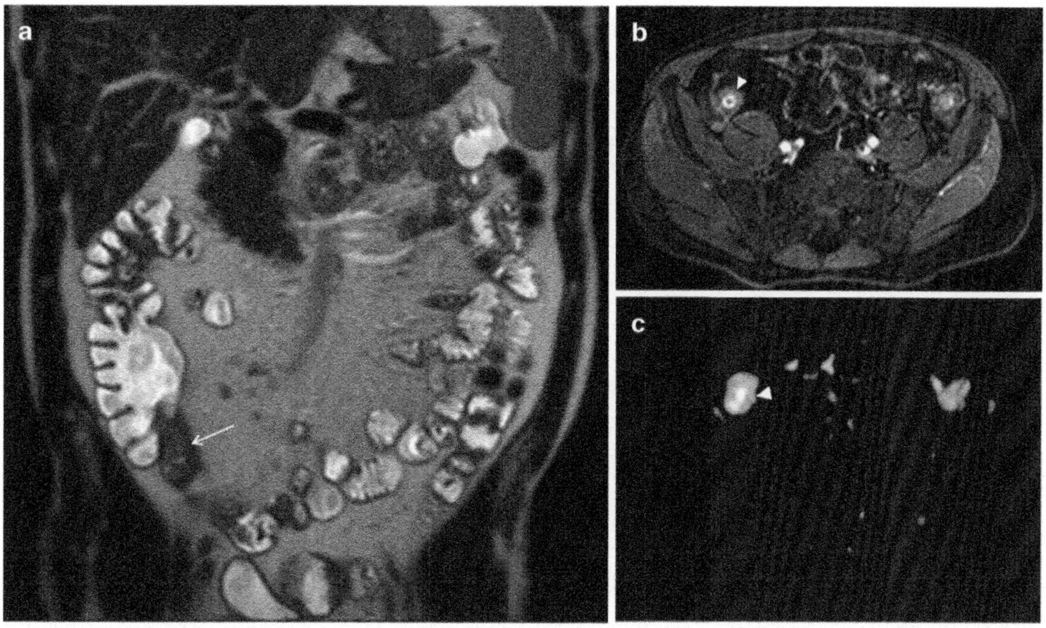

Fig. 11.2 A 51-year-old CD patient with prior ileocecal resection and laterolateral anastomosis. Coronal T2-w HASTE acquisition (**a**) shows severe mural thickening of the ileal anastomotic loop characterized by layered enhancement (arrowhead) on T1-w fat-saturated GE scan (**b**). Inflammation is confirmed by hyperintensity (arrowhead) on axial DWI scan at 800 s/mm^2 b-value (**c**)

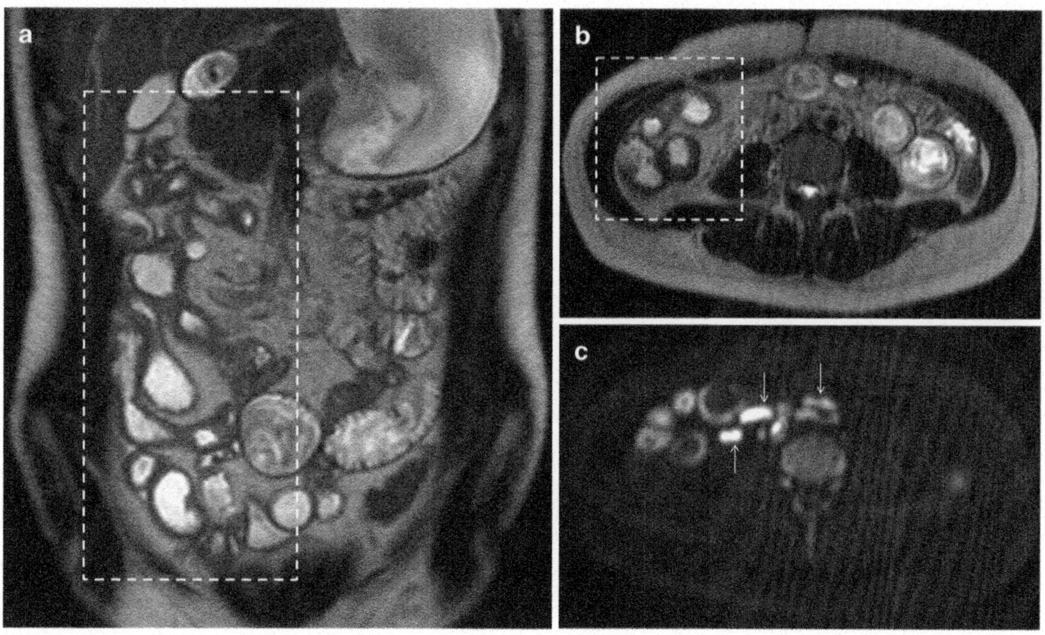

Fig. 11.3 A 22-year-old CD female patient. Coronal (**a**) and axial (**b**) T2-w HASTE images demonstrate mural thickening extensively involving the ileal loops within the right abdominal quadrants (dotted squares). Conversely, jejunal loops, recognizable within the left abdominal quadrants, are spared. Axial DWI scan at 800 s/mm^2 b-value (**c**) shows high signal intensity of the pathologic loops together with mesenteric adenopathies (arrows)

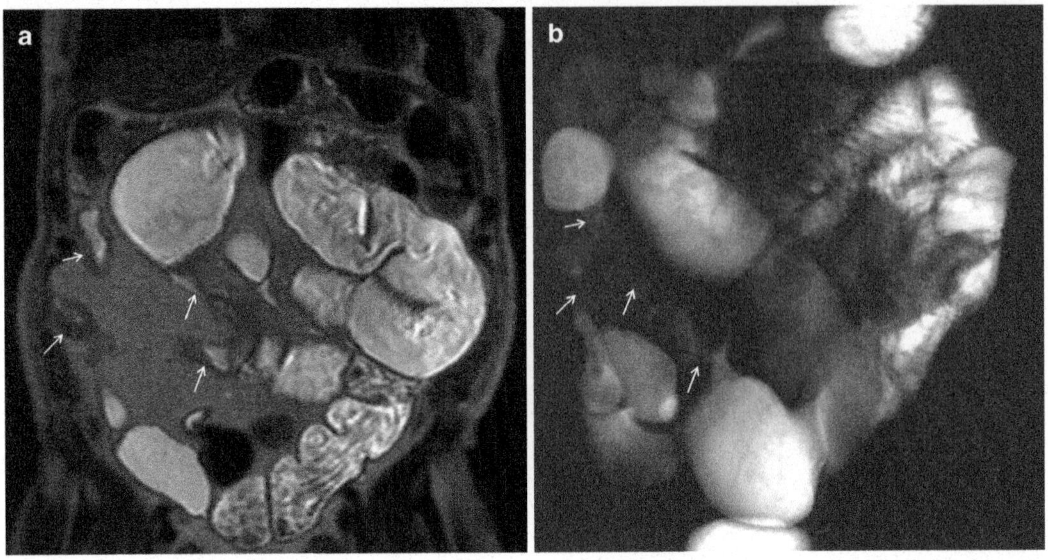

Fig. 11.4 A 65-year-old male patient affected by CD. Multiple wall thickenings, namely "skip lesions" (arrows), are recognizable on coronal T2-w HASTE (**a**) and coronal T2-w RARE thick slab (**b**). Huge dilatation of proximal loops is also visible

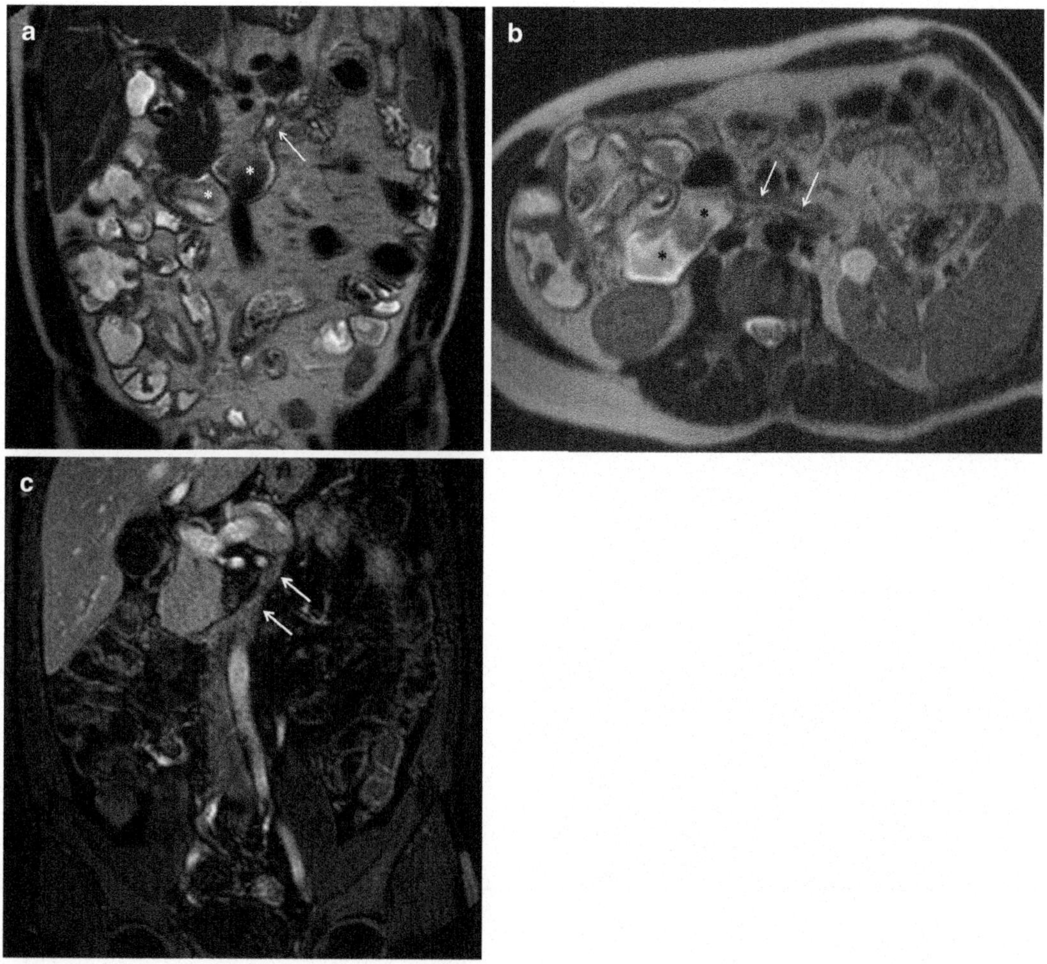

Fig. 11.5 A 48-year-old male patient with CD. Coronal (**a**) and axial-oblique (**b**) T2-w HASTE sequences display a mural thickening of the fourth duodenal portion (arrows) with upstream dilation (asterisks). On coronal GE T1-w scan with fat saturation obtained after Gd injection (**c**) the wall thickening is characterized by slight enhancement

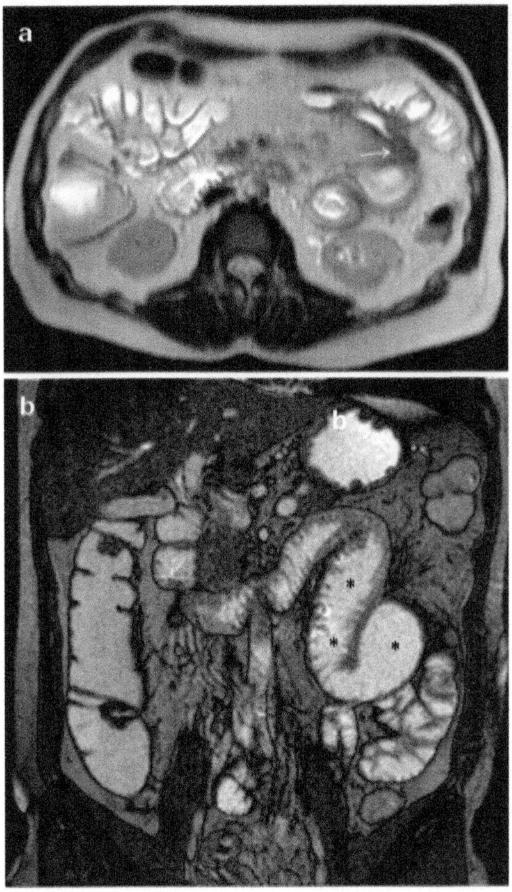

Fig. 11.6 A 47-year-old male CD patient. On axial T2-w HASTE scan (**a**) a jejunal stenosis is visible (arrow). Coronal True-FISP (**b**) shows dilation of the proximal jejunal loops (asterisks) with an increased representation of intestinal folds

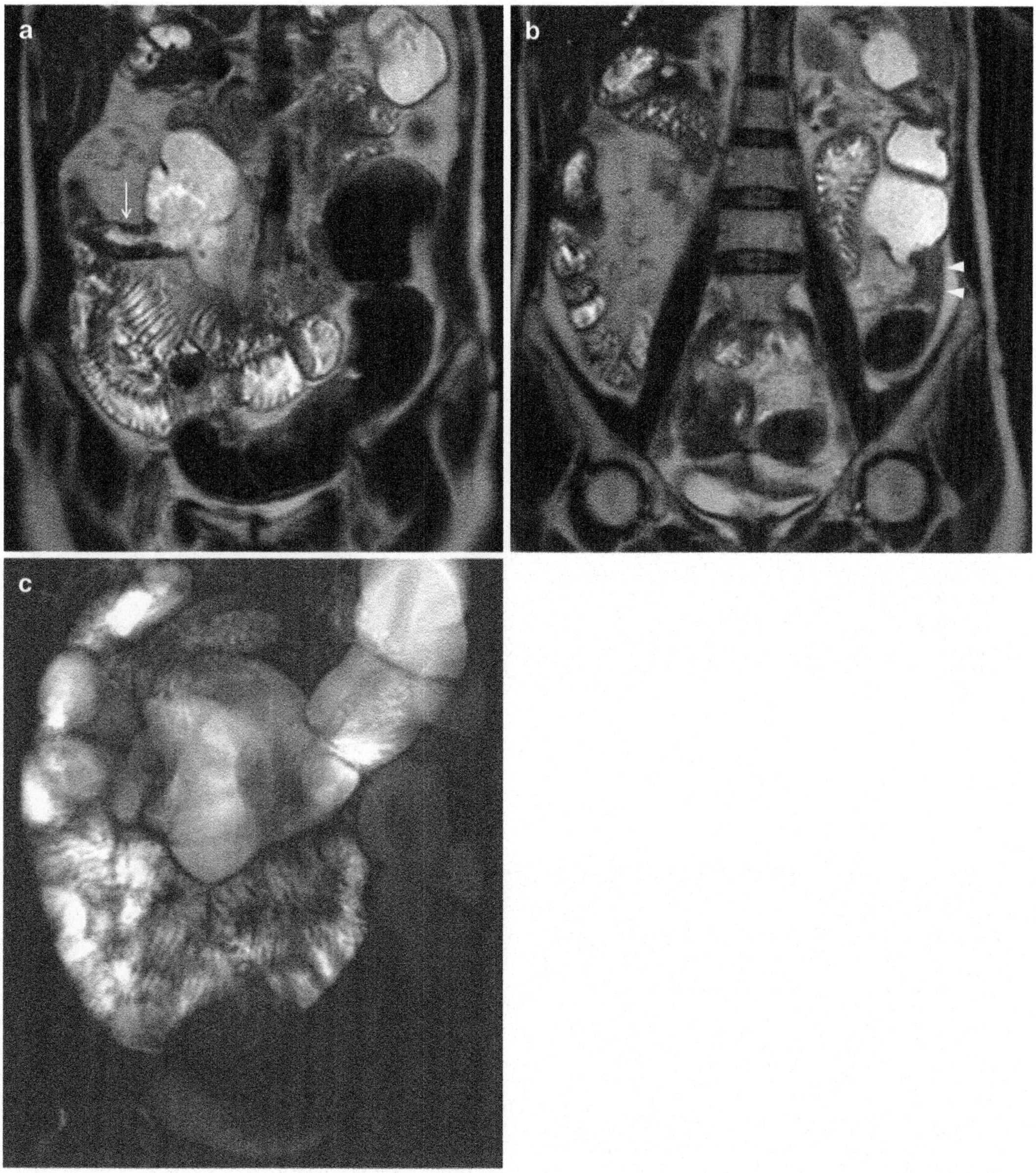

Fig. 11.7 A 49-year-old CD female patient with a prior clinical history of ileocolonic resection and terminoterminal anastomosis construction. Coronal T2-w HASTE image (**a**) shows a mural thickening of the anastomosis (arrow). A more posterior coronal T2-w HASTE scan (**b**) demonstrates another wall thickening within the descending colon (arrowheads). Luminal dilation of small bowel loops with increased number of folds (**c**) is identifiable on coronal T2-w thick slab RARE acquisition

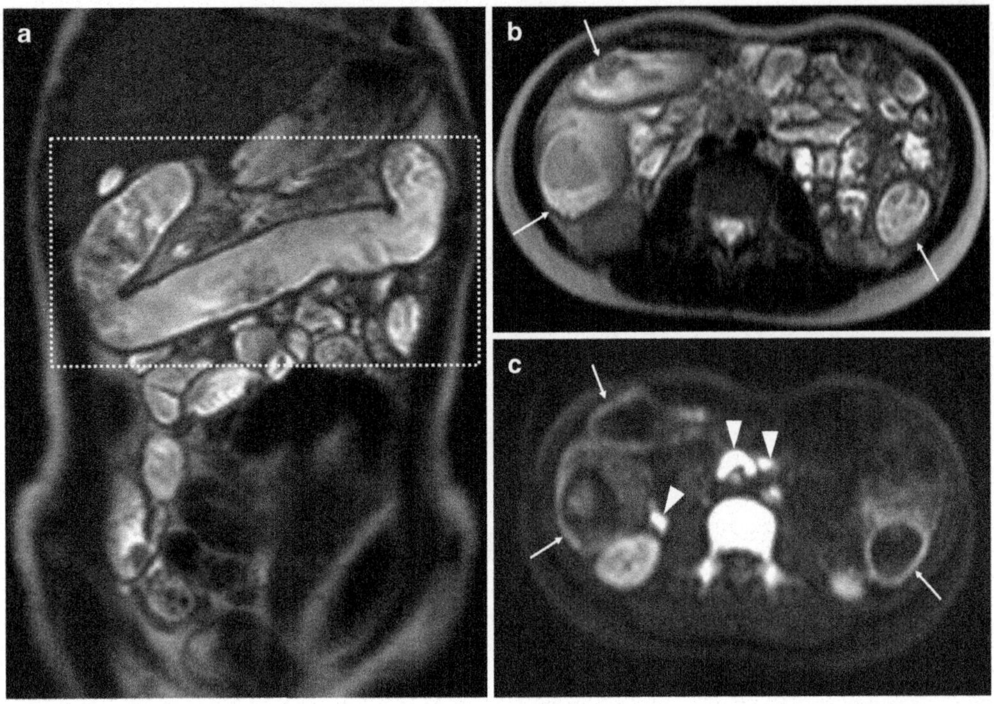

Fig. 11.8 A 9-year-old female CD patient with exclusive involvement of the large bowel. On coronal (**a**) and axial (**b**) T2-w HASTE scans a slight thickening with loss of haustra of the transverse (within the dotted square) as well as of ascending and descending colon (arrows) is appreciable. Axial DWI scan at 800 s/mm^2 b-value (**c**) demonstrates hyperintensity of the colonic walls (arrows) as well as reactive mesenteric lymph nodes (arrowheads)

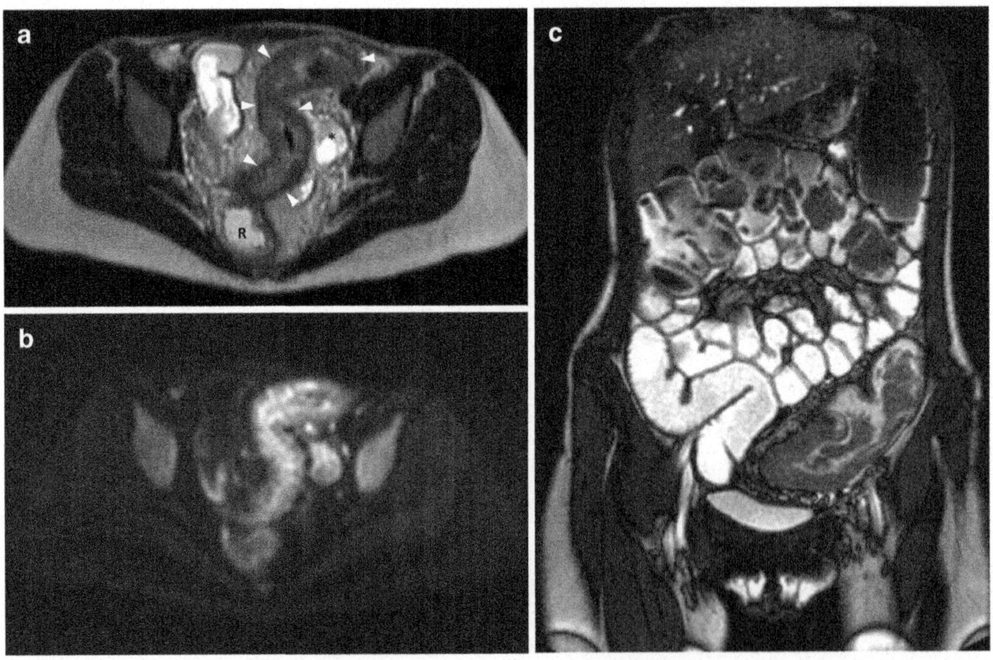

Fig. 11.9 A 21-year-old female patient affected by CD. Extensive concentric wall thickening of the sigmoid colon (arrowheads) is appreciable on axial T2-w HASTE (**a**). Axial DWI scan at 800 s/mm^2 b-value (**b**) demonstrates hyperintensity due to inflammation. Irregularity of the mesenteric border is also visible on coronal True-FISP (**c**) due to vascular engorgement. Asterisk: left ovary; R: rectum

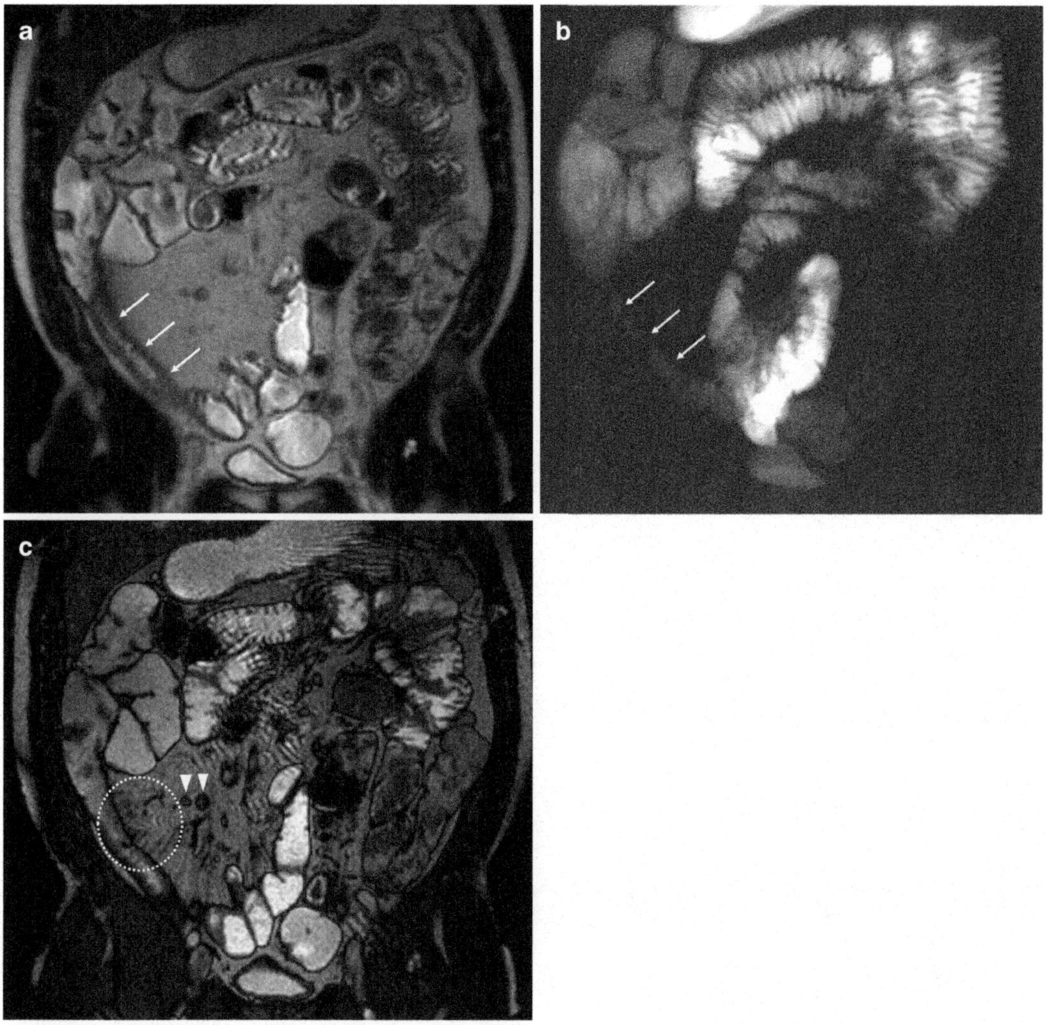

Fig. 11.10 A 54-year-old male CD patient with prior ileocolonic resection and terminoterminal anastomosis construction. Coronal T2-w HASTE (**a**) and coronal T2-w RARE thick slab (**b**) display the wall thickening of the last ileal loop (arrows). Coronal True-FISP (**c**) well demonstrates reactive lymph nodes (arrowheads) and engorgement of vasa recta (within the dotted circle), also known as "comb sign"

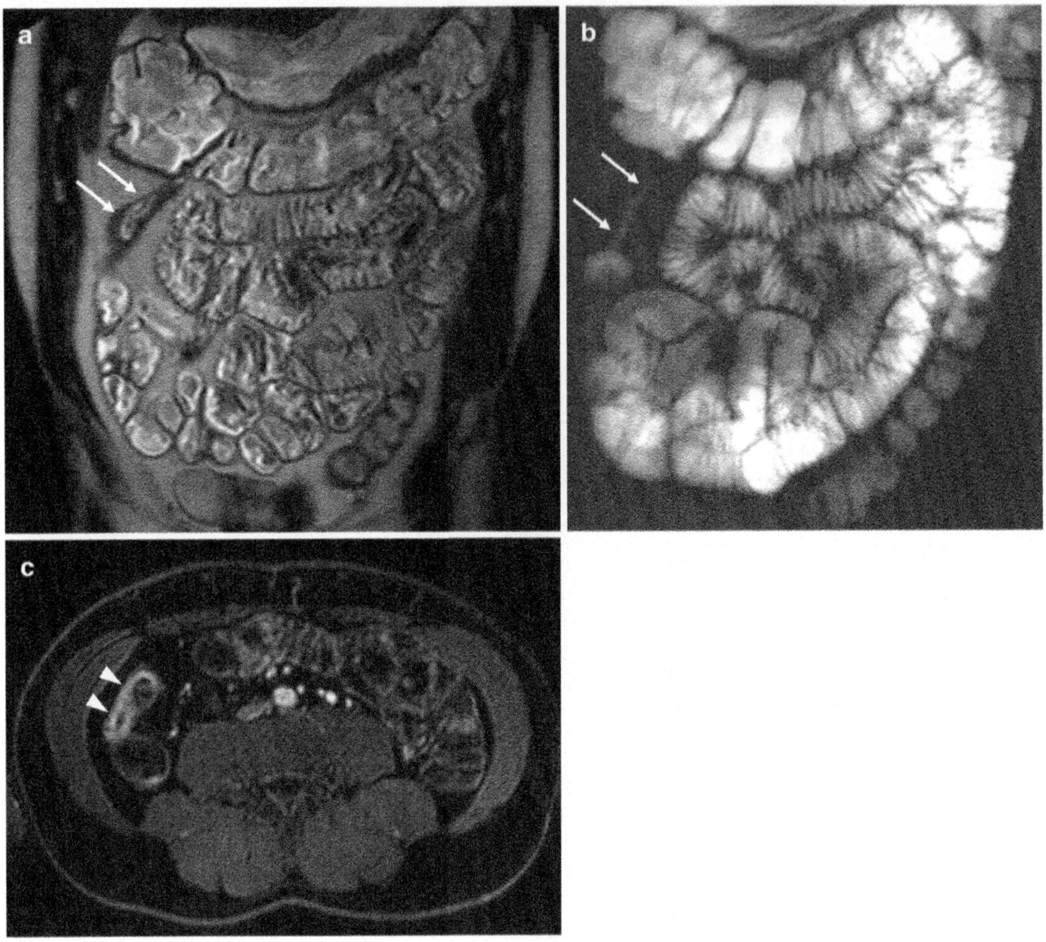

Fig. 11.11 A 44-year-old male CD patient. Coronal T2-w HASTE (**a**) and coronal T2-w RARE thick slab (**b**) images display the outcomes of a previous surgical operation consisting of intestinal resection with ileotransverse anastomosis construction. The last ileal loop is thickened with luminal narrowing (arrows). The diseased intestinal segment shows hyperenhancement (arrowheads) on axial GE T1-w with fat saturation obtained after Gadolinium administration (**c**)

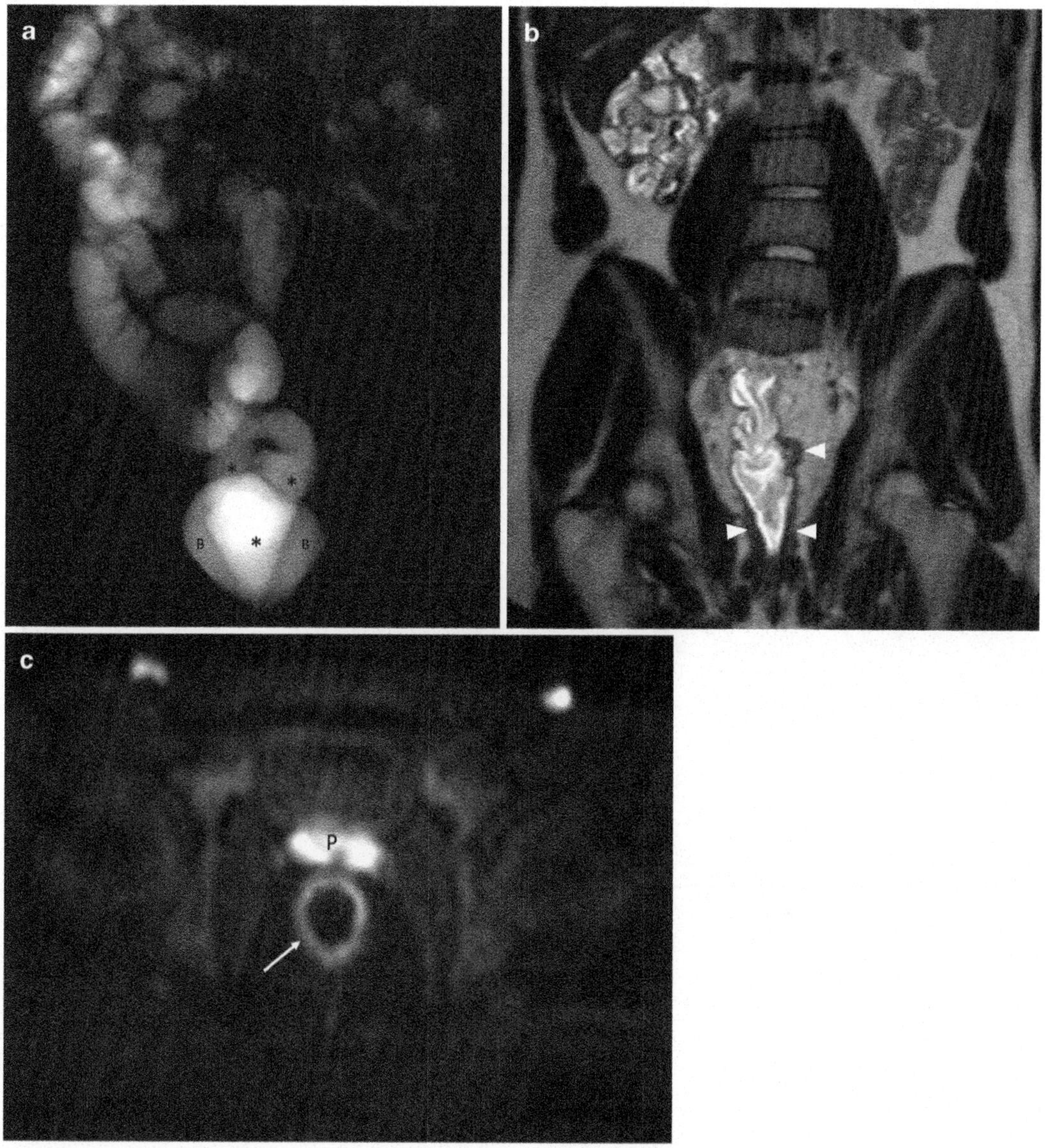

Fig. 11.12 A 17-year-old male UC patient with previous radical colectomy and construction of a J-shaped ileoanal pouch (asterisks) well appreciable on coronal T2-w RARE thick slab (**a**) even with the superimposition of the bladder (b). Coronal T2-w HASTE (**b**) shows a mild thickening of the pouch (arrowheads) due to inflammation ("pouchitis"), as confirmed by the hyperintensity (arrow) on axial DWI scan at 800 s/mm^2 b-value (**c**). P: prostate

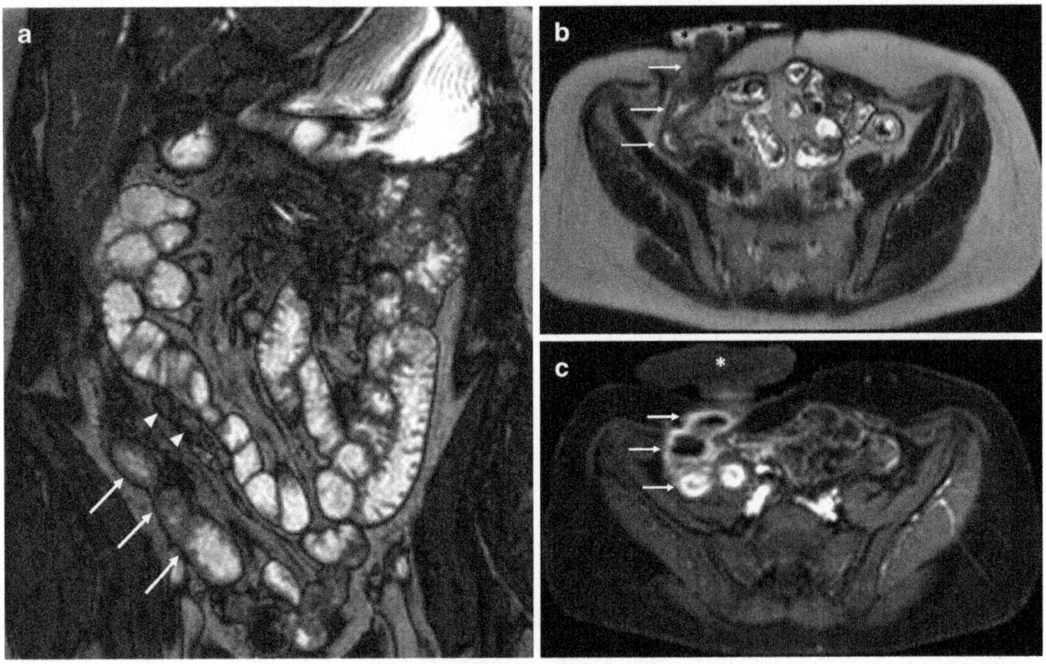

Fig. 11.13 A 52-year-old female CD patient already undergone radical colectomy with oostomy. Coronal True-FISP (**a**) and axial T2-w HASTE (**b**) show an extensive wall thickening of the last ileal loop including the oostomy (arrows). Enlarged reactive lymph nodes are also detectable (arrowheads). Axial GE T1-w scan with fat saturation acquired after Gadolinium injection (**c**) demonstrates hyperenhancement due to active inflammation. Asterisk: external bag for stool collection

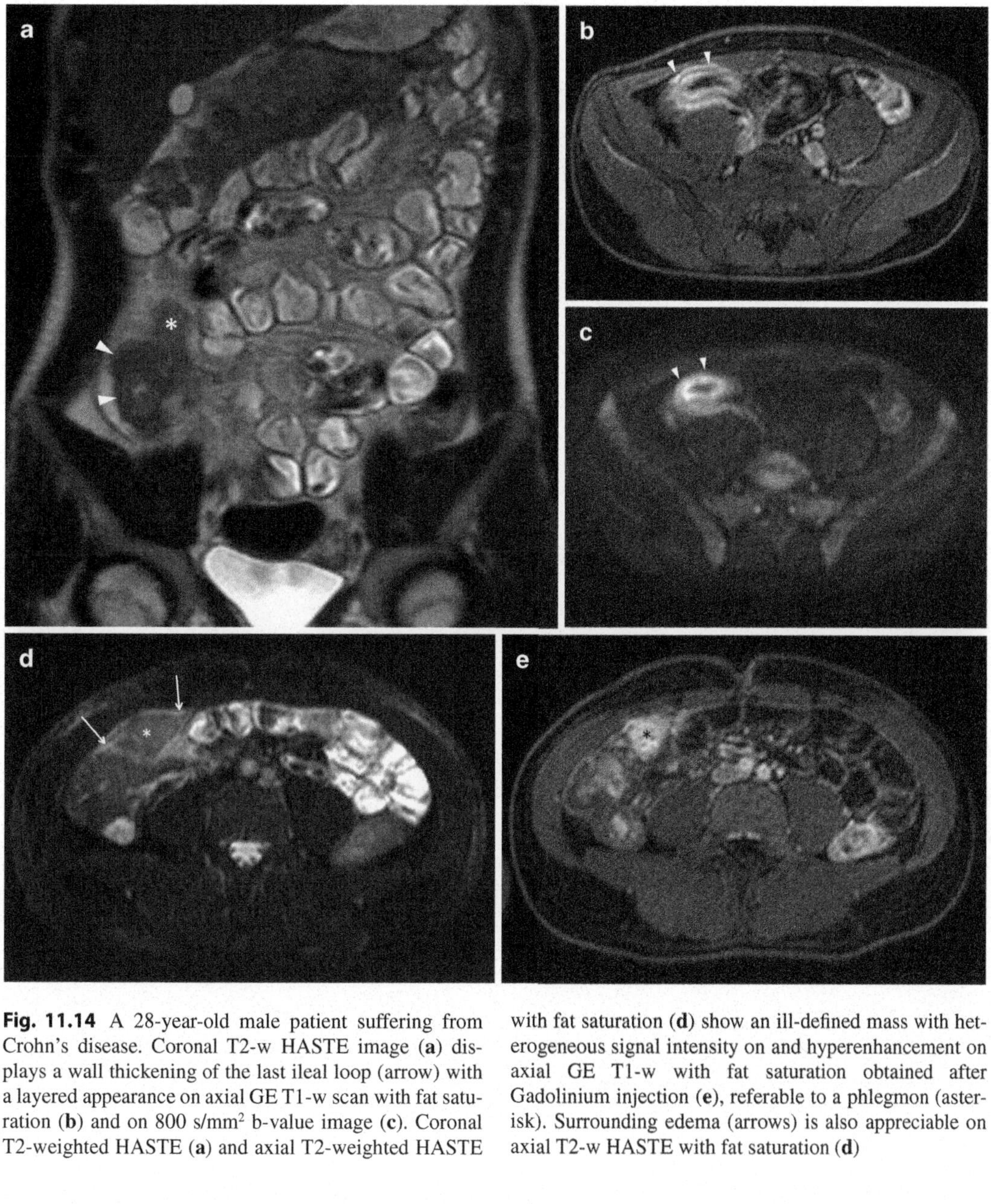

Fig. 11.14 A 28-year-old male patient suffering from Crohn's disease. Coronal T2-w HASTE image (**a**) displays a wall thickening of the last ileal loop (arrow) with a layered appearance on axial GE T1-w scan with fat saturation (**b**) and on 800 s/mm^2 b-value image (**c**). Coronal T2-weighted HASTE (**a**) and axial T2-weighted HASTE with fat saturation (**d**) show an ill-defined mass with heterogeneous signal intensity on and hyperenhancement on axial GE T1-w with fat saturation obtained after Gadolinium injection (**e**), referable to a phlegmon (asterisk). Surrounding edema (arrows) is also appreciable on axial T2-w HASTE with fat saturation (**d**)

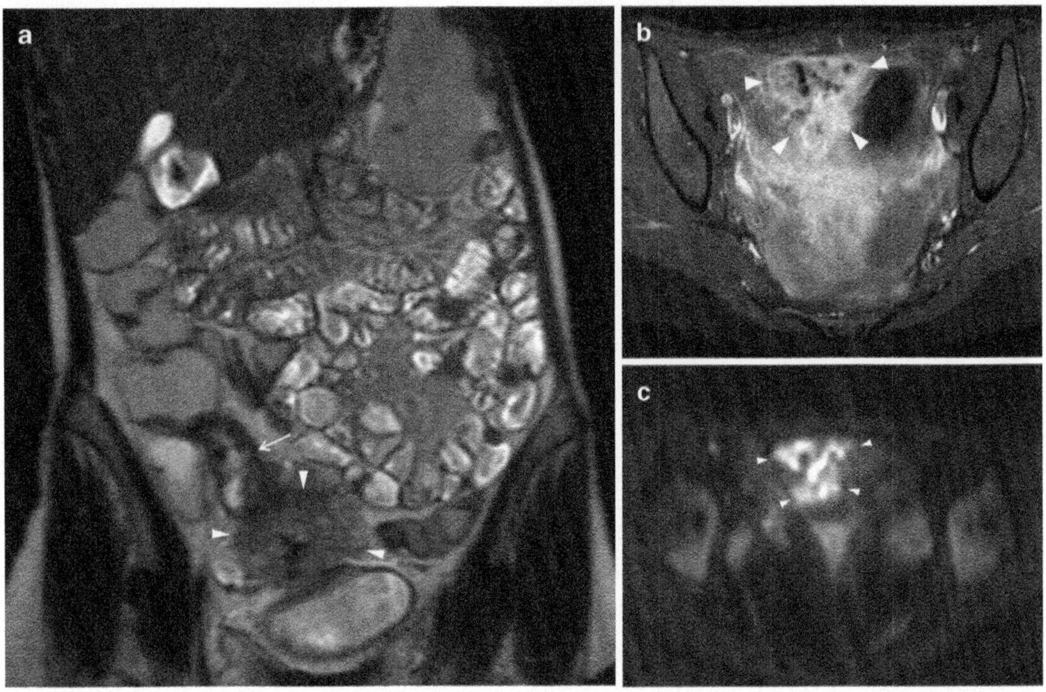

Fig. 11.15 A 22-year-old CD female patient. Coronal T2-w HASTE sequence (**a**) shows a thickened terminal ileum (arrow) with an abscessual collection (arrowheads). Axial T1-w fat-saturated image obtained after Gadolinium injection (**b**) demonstrates inhomogeneous enhancement of the collection with small signal voids due to gas bubbles. The axial DWI scan at 800 s/mm^2 b-value (**c**) further confirms the presence of an abscess

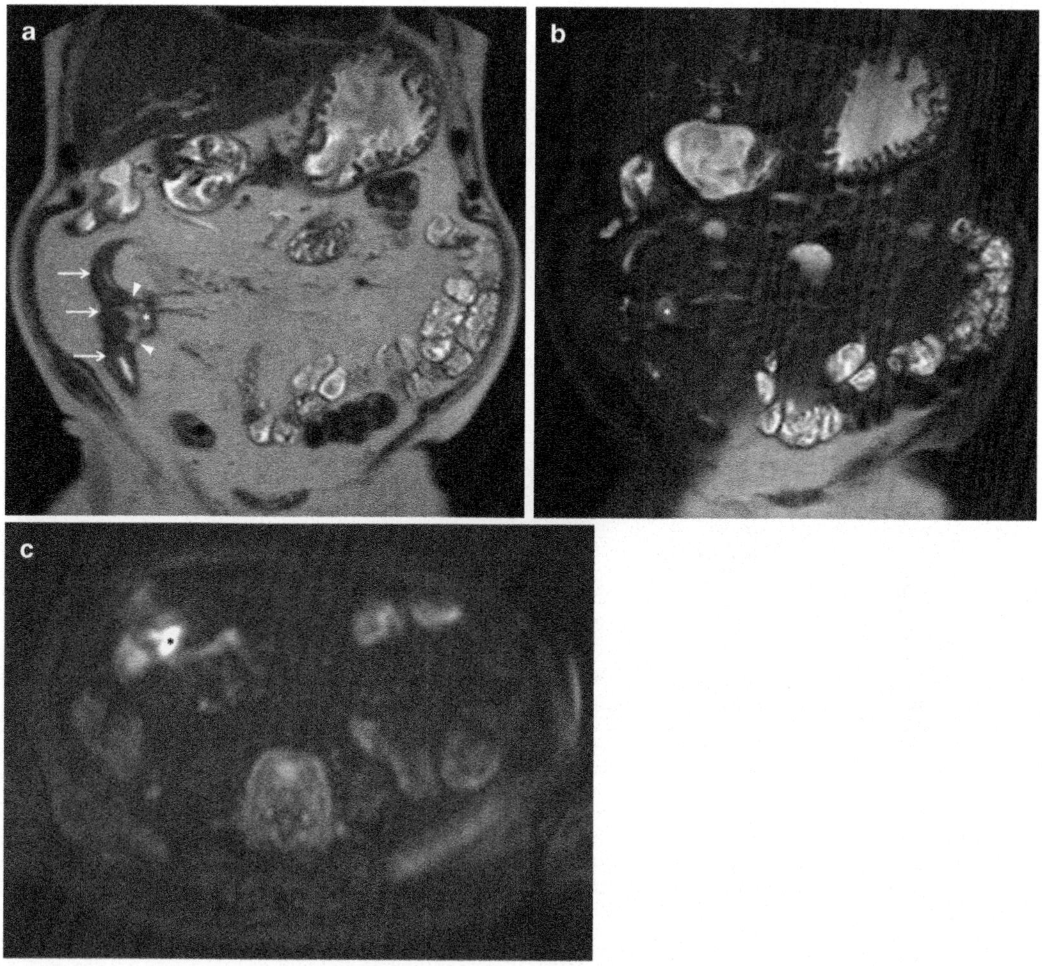

Fig. 11.16 A 65-year-old male CD patient already undergone intestinal surgery. Coronal T2-w HASTE without (**a**) and with fat saturation (**b**) show mural thickening of an ileal loop (arrows) within the right abdominal quadrant. Fistulous tracts (arrowheads) connect the ileal loop to a small abscessual collection (asterisk) characterized by hyperintensity on axial DWI scan at 800 s/mm^2 b-value (**c**)

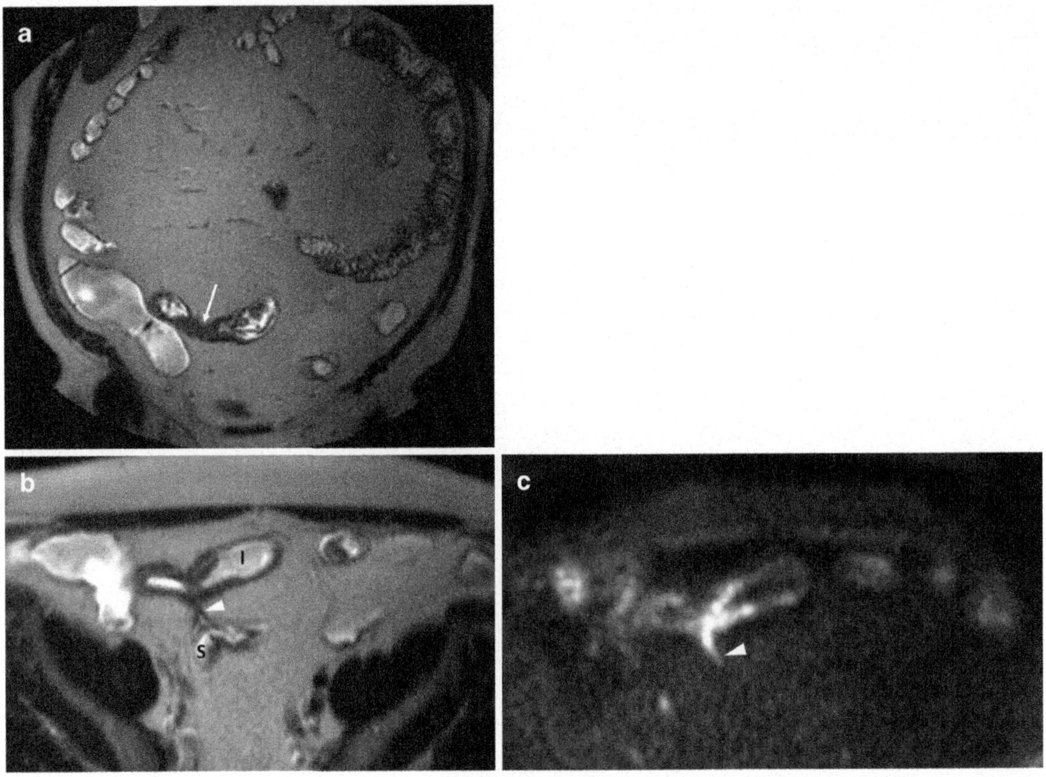

Fig. 11.17 A 51-year-old male CD patient. Coronal T2-w HASTE scan (**a**) shows a thickened ileal loop with luminal narrowing (arrow). Magnified axial T2-w HASTE image (**b**) and axial DWI at 800 s/mm^2 b-value scan (**c**) demonstrate a fistulous tract (arrowhead) connecting the ileum (I) to the sigmoid colon (S)

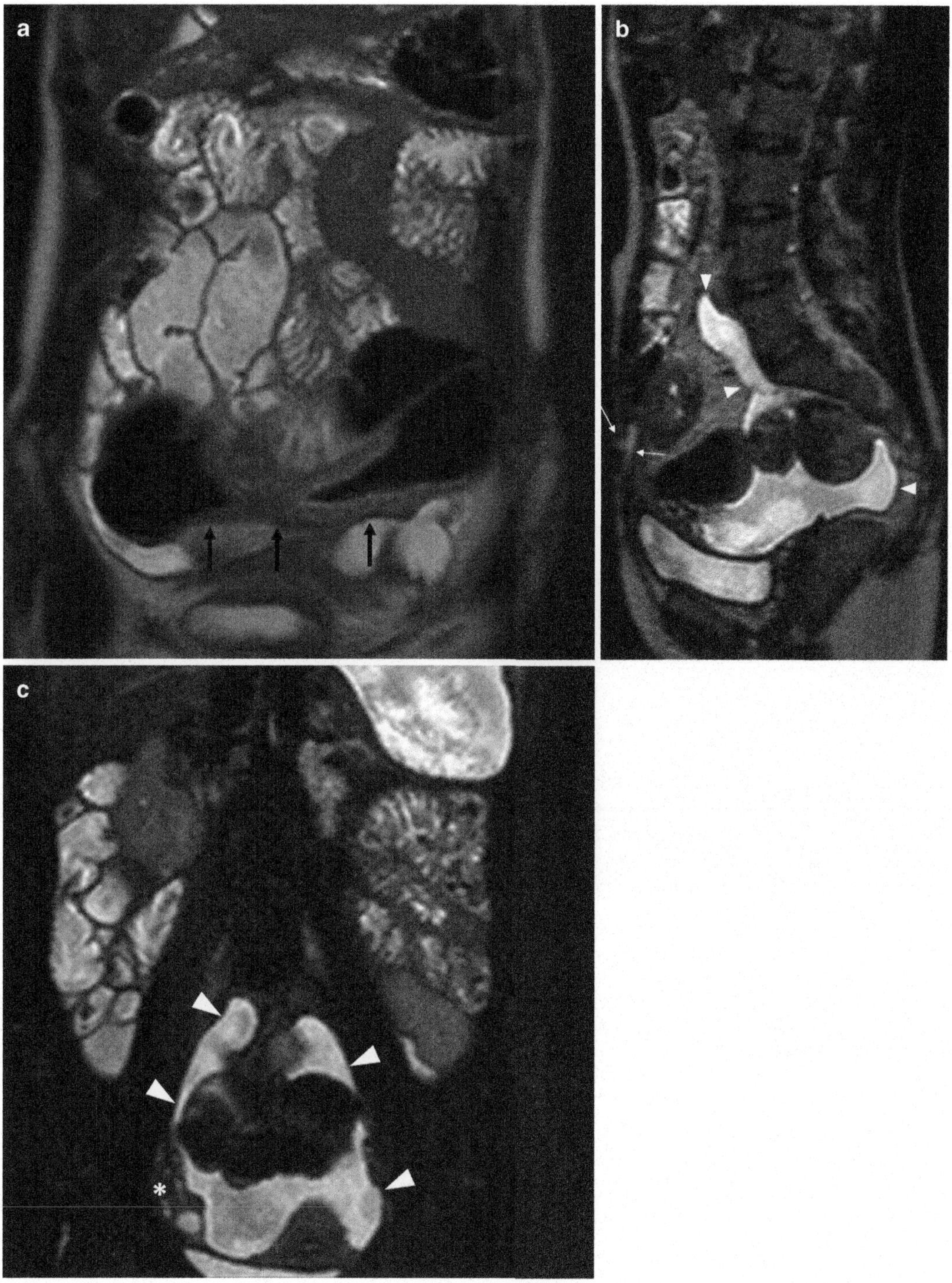

Fig. 11.18 A 35-year-old female CD patient with previous surgical intervention. On coronal T2-w HASTE (**a**) a thickened ileal loop with a stricture is detectable (black arrows). Sagittal T2-w HASTE (**b**) further shows an enterocutaneous fistula (white arrows). On coronal T2-w HASTE scan with fat saturation (**c**) is better identifiable a fluid collection (arrowheads) in which the ovaries are included (asterisk: right ovary) referable to a peritoneal inclusion cyst

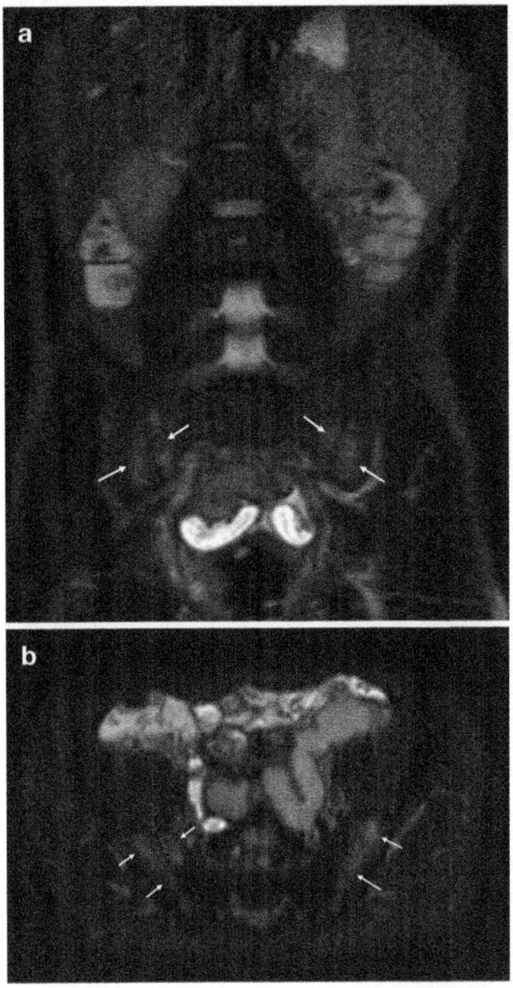

Fig. 11.19 A 35-year-old CD male patient. MRE was unremarkable for intestinal findings. On the other hand, coronal (**a**) and axial (**b**) T2-w HASTE scans with fat saturation demonstrated hyperintensity of the sacroiliac joints on both sides (arrows) due to edema. The findings are consistent with sacroileitis

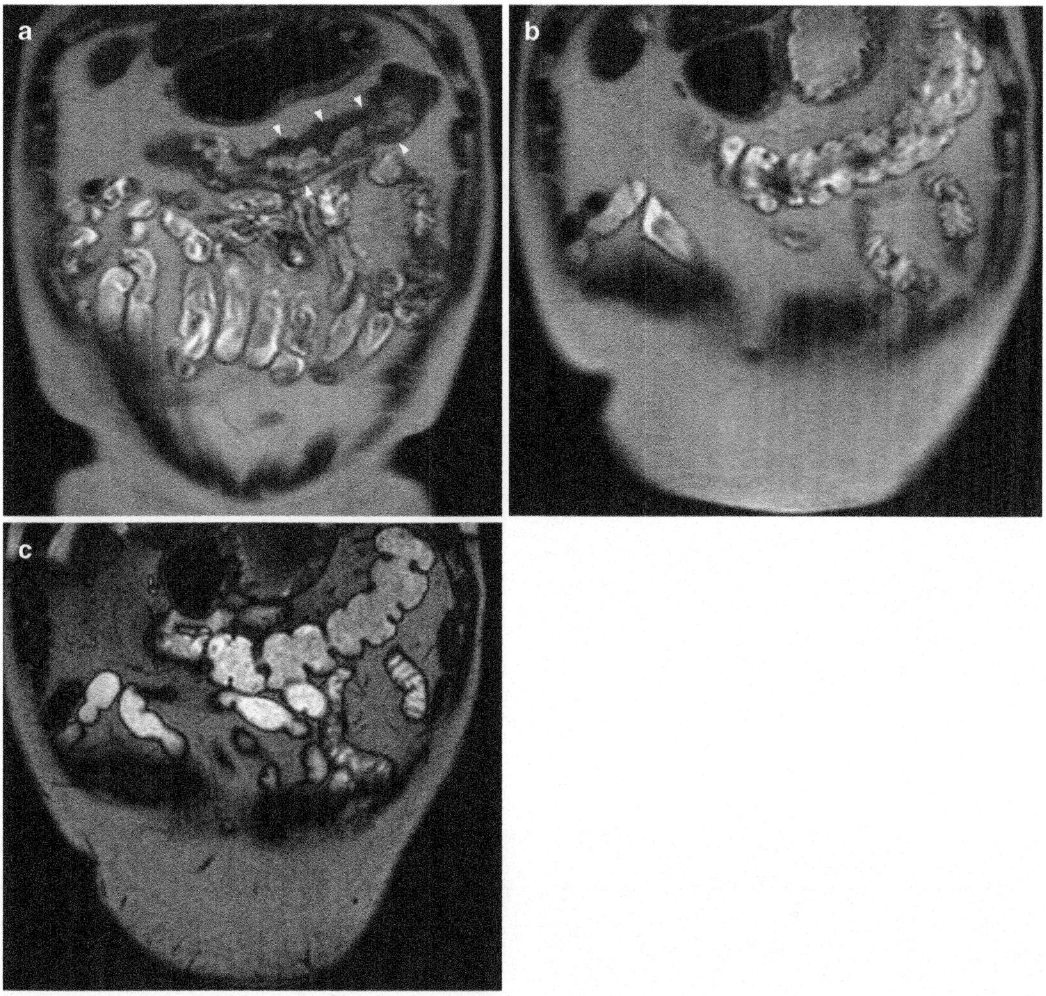

Fig. 11.20 A 77-year-old male patient with clinical suspicion of IBD. On coronal T2-w HASTE (**a**) a wall thickening of the left transverse hemicolon seems to be present (arrowheads). Delayed coronal T2-w HASTE (**b**) and coronal True-FISP (**c**) images performed in prone position demonstrate normal mural edges. The initial finding was a pitfall caused by poor luminal dilation

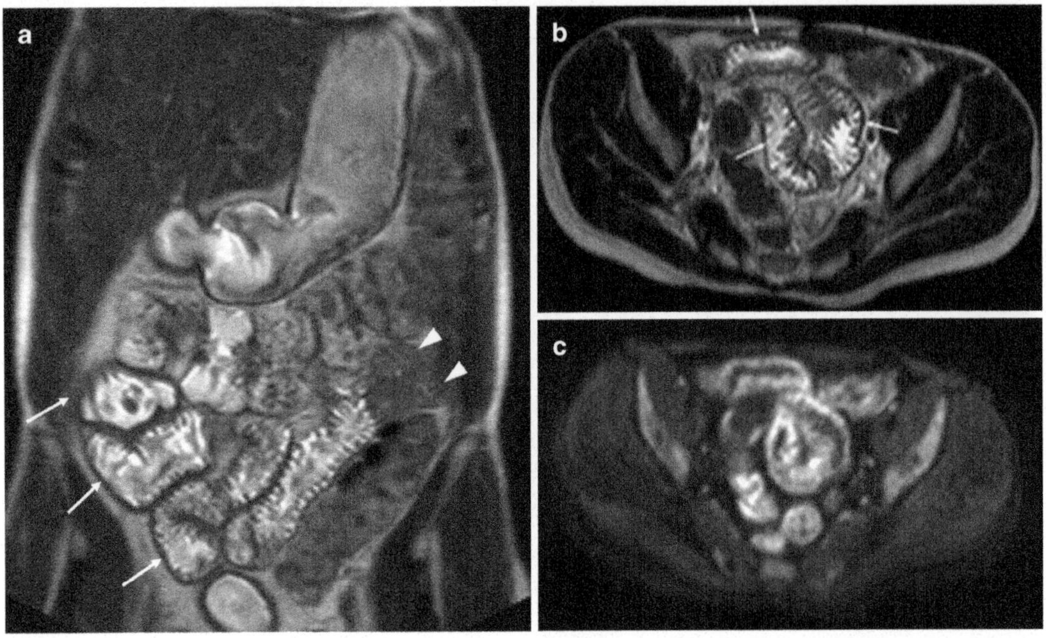

Fig. 11.21 A 53-year-old male patient suffering from celiac disease. Coronal (**a**) and axial (**b**) T2-w HASTE sequences show an increase in number and thickness of folds of the ileal loops (arrows) also known as "ileal jejunization." On the contrary, jejunal loops (arrowheads) are collapsed. Axial DWI scan at 800 s/mm^2 b-value (**c**) shows a diffuse hyperintensity of the ileal loops

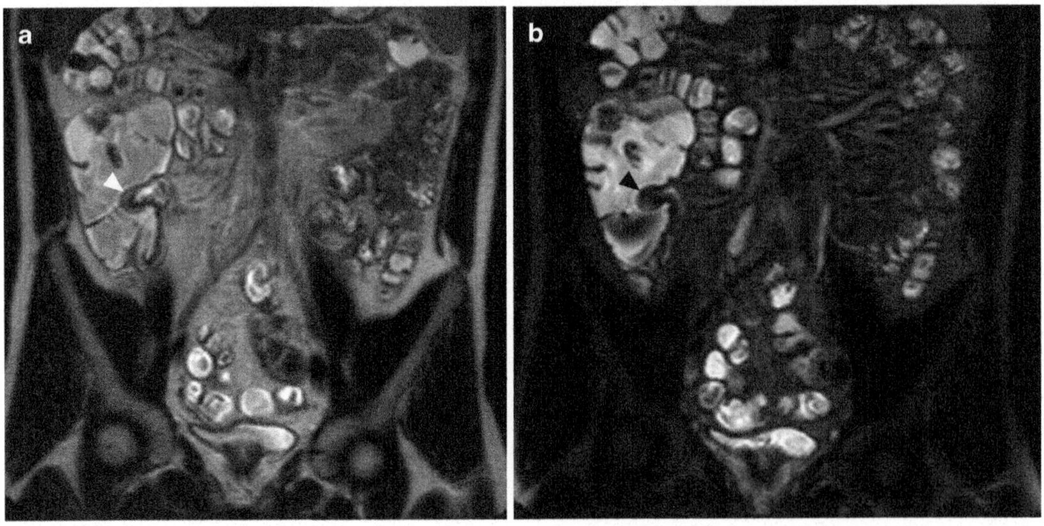

Fig. 11.22 A 28-year-old male patient was hospitalized due to recurrent episodes of intestinal subocclusion. Coronal T2-w HASTE scan (**a**) shows a mild thickening of the ileocecal valve (arrowhead) with drop of signal intensity on coronal T2-w HASTE with fat saturation (**b**) referable to lipomatosis. The diagnosis was later confirmed by endoscopy

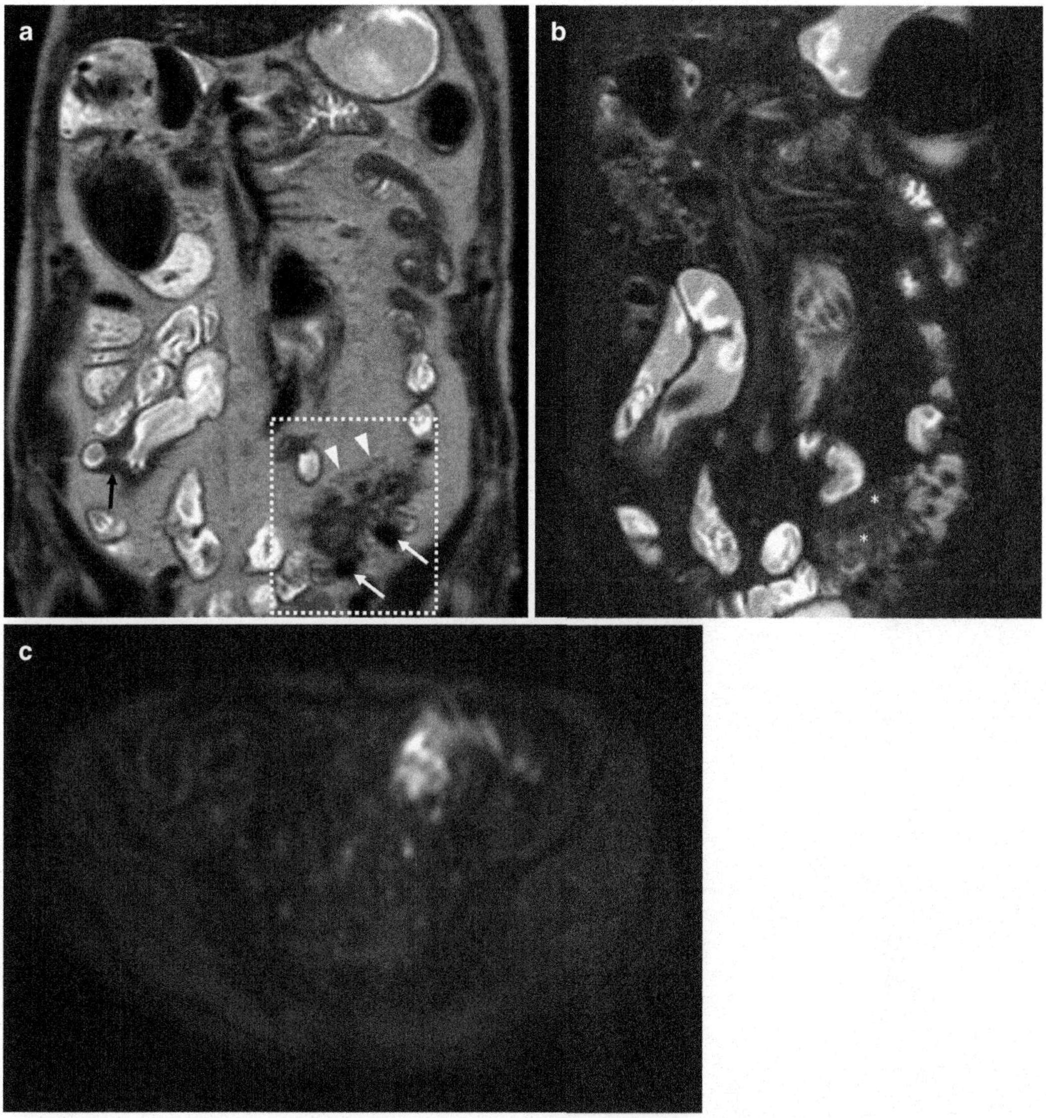

Fig. 11.23 A 85-year-old male patient with a known diagnosis of CD was hospitalized due to persistent abdominal pain. On Coronal T2-w HASTE scan (**a**) a mural thickening of an ileal loop within the right abdominal quadrant (black arrow) is detectable and consistent with a CD lesion. However, the same image also shows an irregular wall thickening of the sigmoid colon within the left iliac fossa (dotted square). Signal voids due to gas inside sigmoid diverticula (white arrows) and vascular engorgement (arrowheads) are also present. The sigmoid walls as well as the surrounding adipose tissue are further characterized by hyperintensity on T2-w HASTE with fat saturation scan (**b**), due to edema (asterisks), and on axial DWI scan at 800 s/mm^2 b-value (**c**), related to inflammation. The final diagnosis was acute diverticulitis

11.2 MRI Perianal Involvement Cases (Figs. 11.24, 11.25, 11.26, 11.27, 11.28, 11.29 and 11.30)

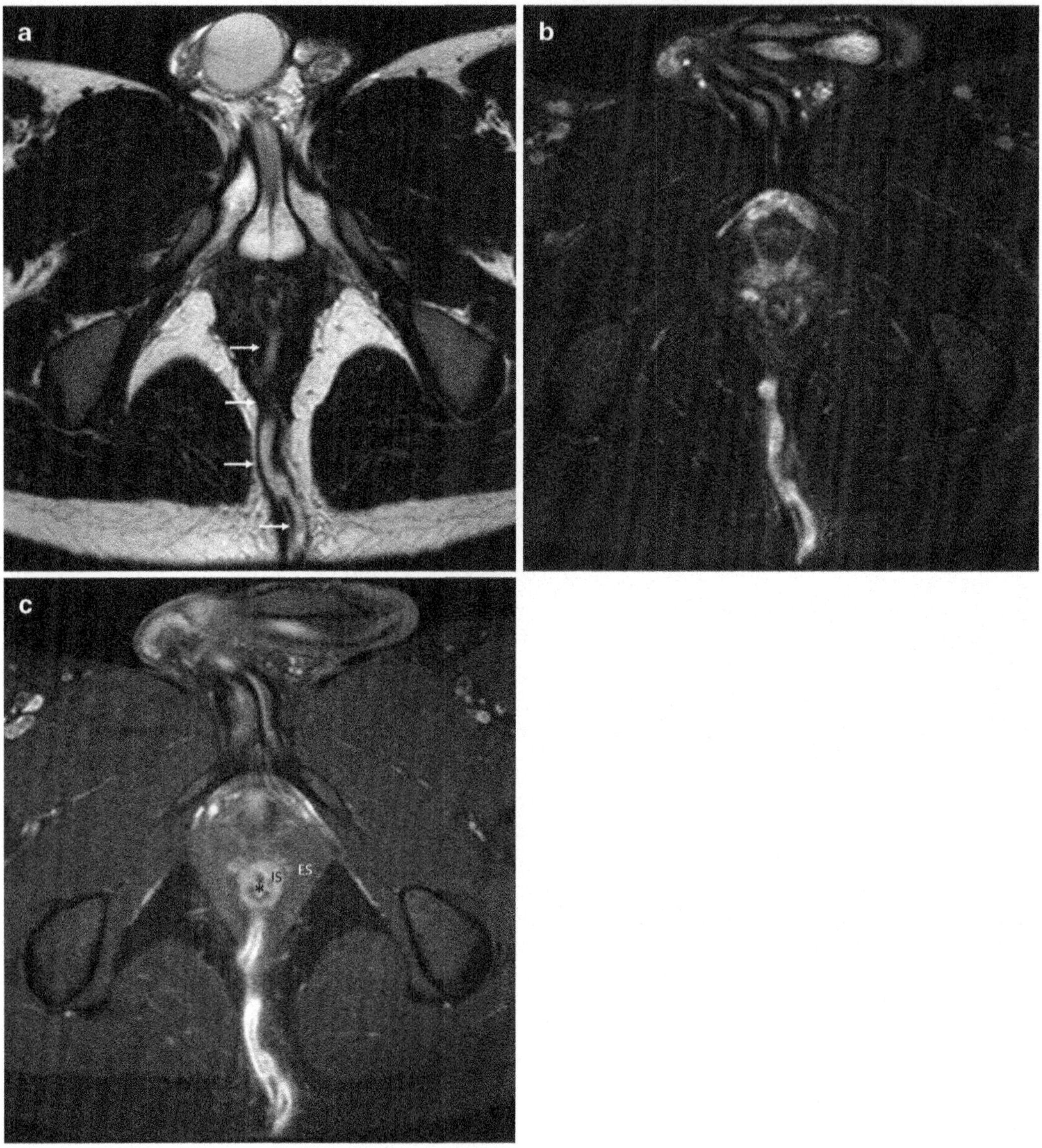

Fig. 11.24 A 21-year-old CD male patient. Axial-oblique T2-w TSE (**a**) shows a trans-sphincteric fistula (arrows) extending from the anal canal to the cutaneous surface. The fistulous tract is hyperintense on axial-oblique T2-w SPAIR (**b**) and axial-oblique GE T1-w with fat saturation after Gadolinium injection (**c**). Asterisk: anal lumen; *IS* internal sphincter; *ES* external sphincter

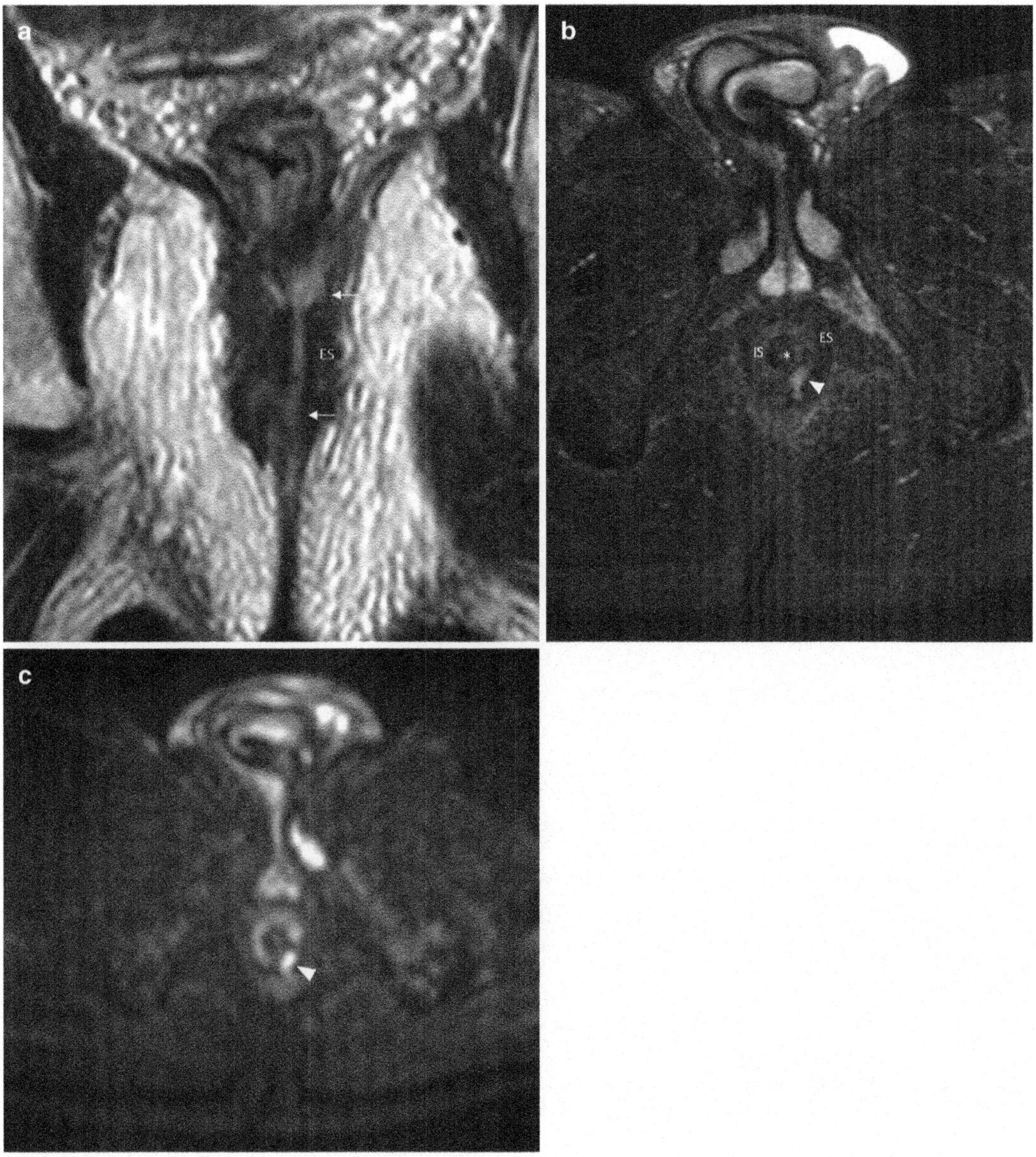

Fig. 11.25 A 54-year-old male CD patient. Coronal-oblique T2-w TSE scan (**a**) displays an inter-sphincteric fistula (arrows). Axial-oblique T2-w SPAIR (**b**) and axial-oblique DWI scan at 800 s/mm^2 b-value (**c**) well demonstrate the origin of the fistula (arrowhead) at the left posterolateral side of the anal canal (4–5 o'clock according to the "anal clock"). Asterisk: anal lumen; *IS* internal sphincter, *ES* external sphincter

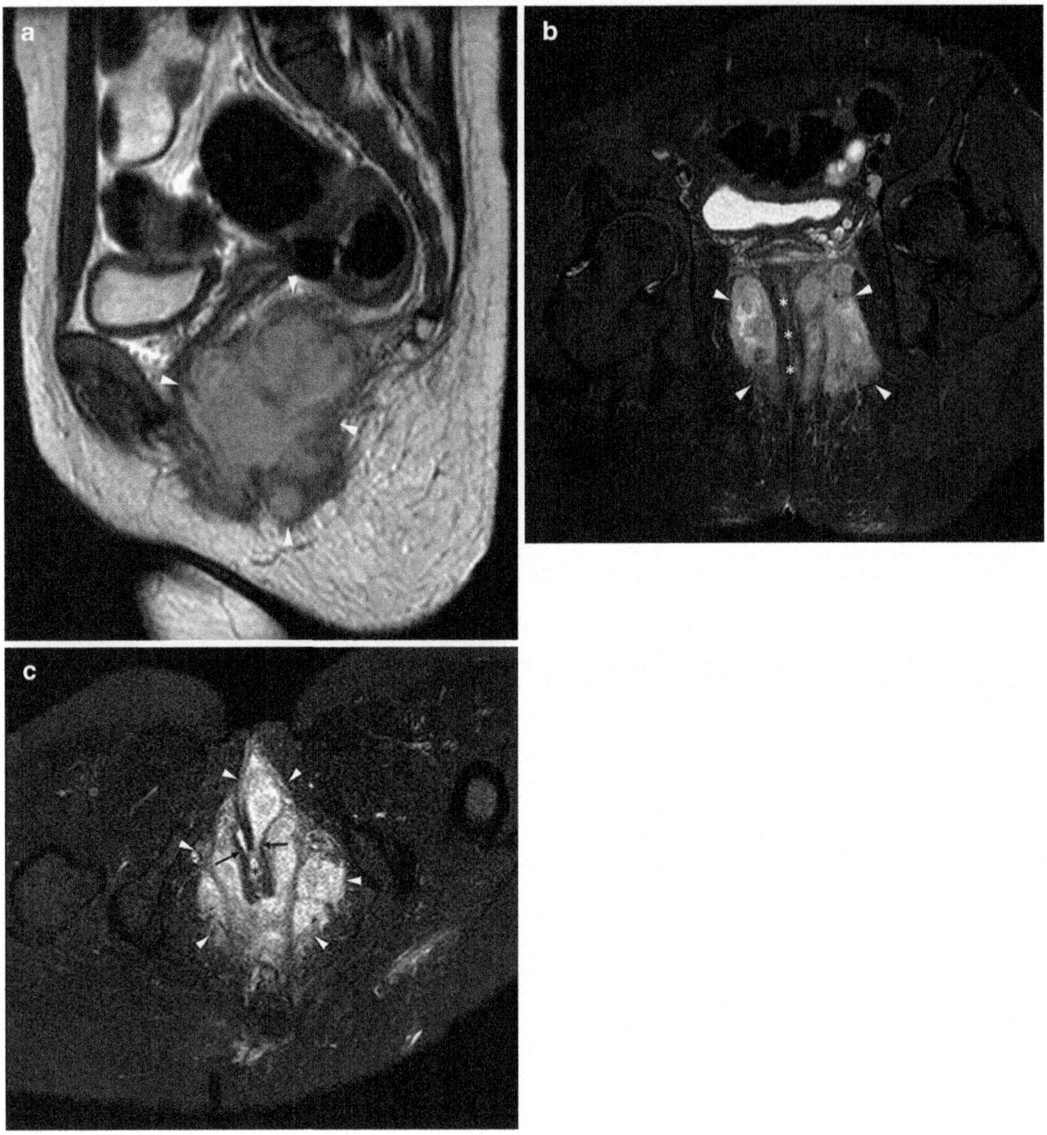

Fig. 11.26 A 22-year-old female CD patient. On sagittal T2-w TSE (**a**) and coronal-oblique T2-w SPAIR (**b**) a large abscess, extending into both the ischioanal fossae, is visible (arrowheads). Axial-oblique T2-w SPAIR (**c**) demonstrates two fistulous tracts (arrows) connecting the abscess to the anal canal (asterisks)

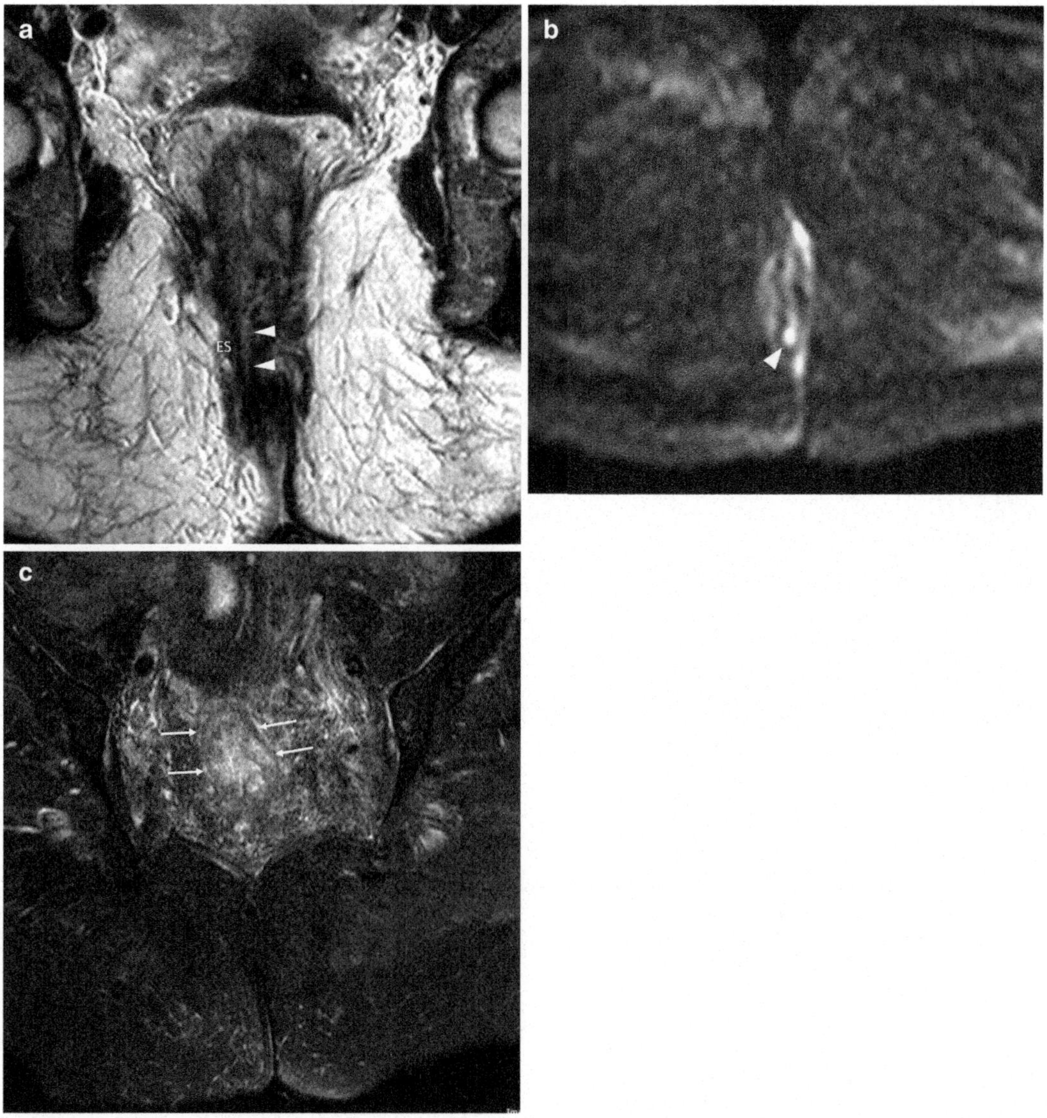

Fig. 11.27 A 50-year-old female CD patient. On coronal-oblique T2-w TSE image (**a**) an inter-sphincteric fistula (arrowheads), characterized by high signal intensity on axial-oblique DWI at 800 s/mm^2 b-value (**b**). Coronal-oblique T2-w SPAIR (**c**) shows hyperintensity of the rectal walls (arrows) due to edema and referable to proctitis (arrows)

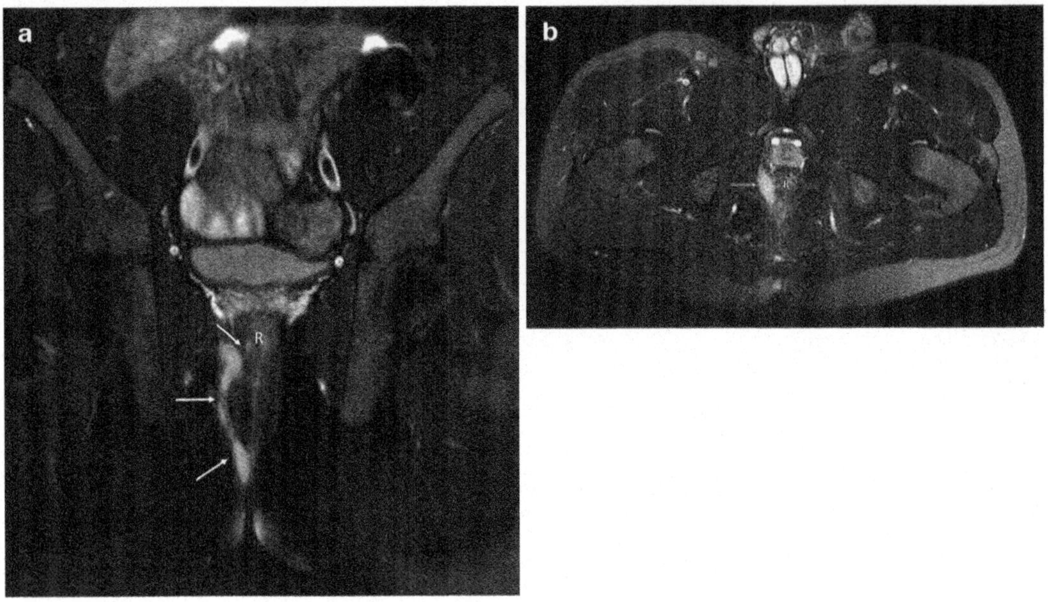

Fig. 11.28 A 45-year-old CD male patient. Coronal-oblique T2-w SPAIR (**a**) demonstrates a fistulous tract (arrows) originating from the rectum (R), passing through the right levator ani muscle, and extending inferiorly till the cutaneous surface. Axial-oblique T2-w SPAIR (**b**) shows the origin of the fistula (arrow) from the right wall of the rectum (R)

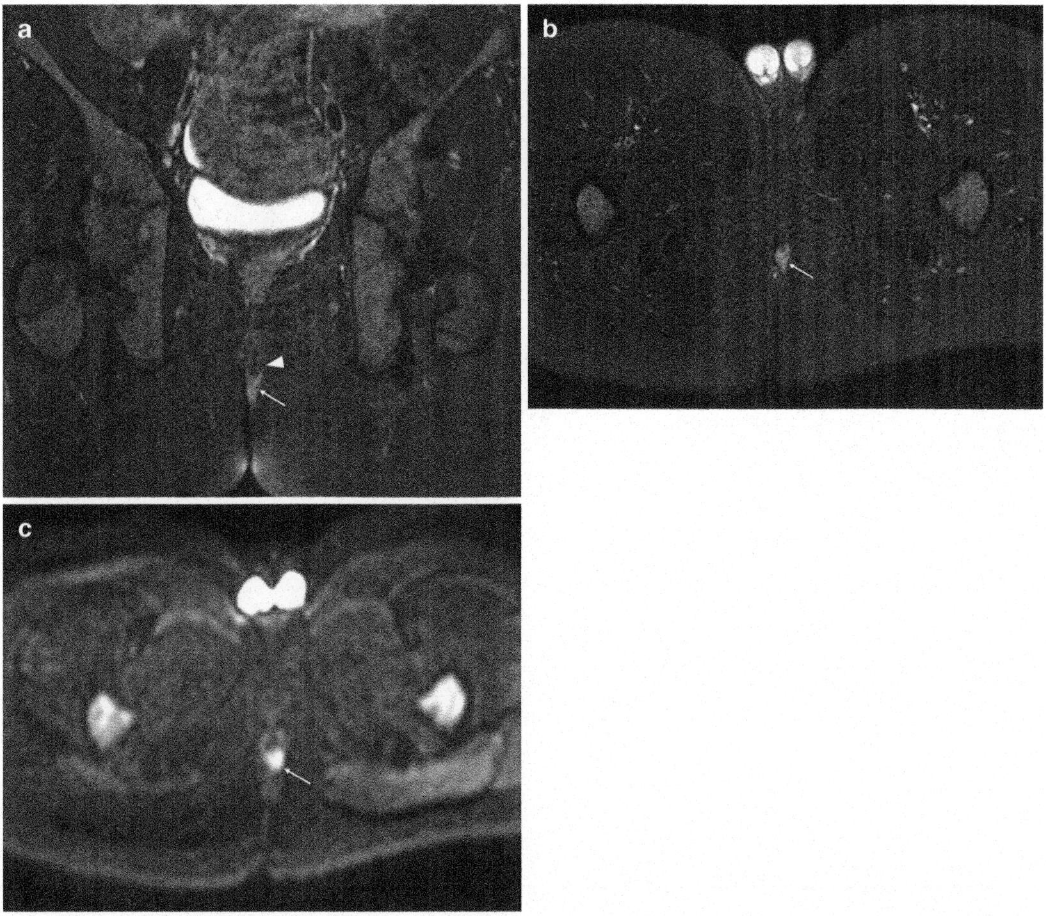

Fig. 11.29 A 13-year-old CD male patient. Coronal-oblique (**a**) and axial-oblique (**b**) T2-w SPAIR, as well as axial-oblique DWI at 800 s/mm^2 b-value (**c**), are capable in showing an hyperintense fistula (arrows) adjacent to the left external sphincter (arrowhead). In this case, the fistula does not traverse any sphincter and can be classified as "superficial"

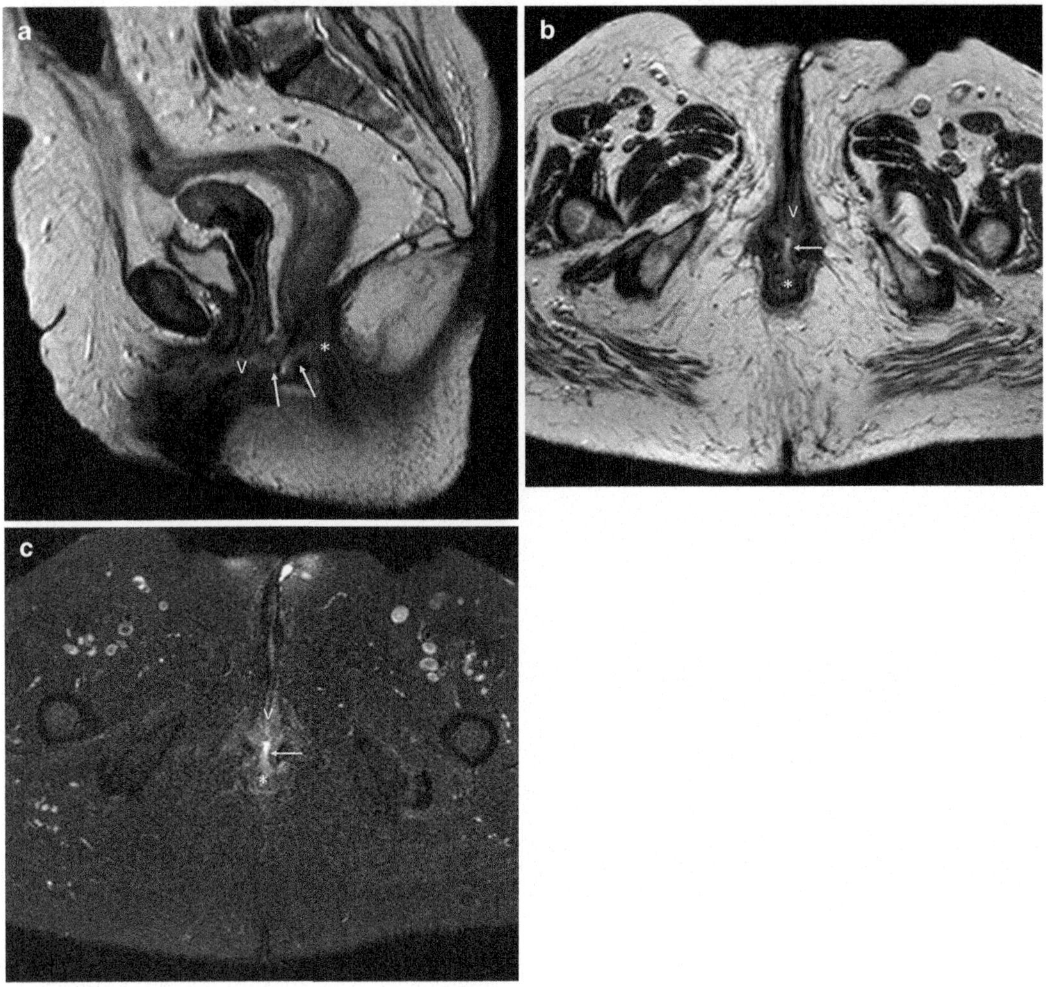

Fig. 11.30 A 73-year-old female CD patient. Sagittal (**a**) and axial-oblique (**b**) T2-w TSE as well as axial-oblique T2-w SPAIR (**c**) display a fistula (arrows) connecting the anal canal (asterisk) with the vaginal lumen (V)

SPRINGER NATURE

GPSR Compliance

The European Union's (EU) General Product Safety Regulation (GPSR) is a set of rules that requires consumer products to be safe and our obligations to ensure this.

If you have any concerns about our products, you can contact us on ProductSafety@springernature.com

In case Publisher is established outside the EU, the EU authorized representative is:

Springer Nature Customer Service Center GmbH
Europaplatz 3
69115 Heidelberg, Germany

The manufacturer's authorised representative in the EU is Springer Nature Customer Service Centre GmbH, Europaplatz 3, 69115 Heidelberg, Germany. If you have any concerns regarding our products, please contact ProductSafety@springernature.com

Printed and bound by CPI Group (UK) Ltd, Croydon, CR0 4YY

31/07/2026

02182834-0002